Algorithms and Combinatorics 17

Springer
Berlin
Heidelberg
New York
Barcelona
Hong Kong
London
Milan
Paris
Singapore
Tokyo

Oded Goldreich

Modern Cryptography, Probabilistic Proofs and Pseudorandomness

 Springer

Oded Goldreich
Department of Computer Science
and Applied Mathematics
The Weizmann Institute of Science
76100 Rehovot
Israel
e-mail: oded@wisdom.weizmann.ac.il

Cataloging-in-Publication Data applied for

Die Deutsche Bibliothek – CIP-Einheitsaufnahme

Goldreich, Oded: Modern cryptography, probabilistic proofs and pseudorandomness / Oded Goldreich. – Berlin; Heidelberg; New York; Barcelona; Hong Kong; London; Milan; Paris; Singapore; Tokyo: Springer, 1999
(Algorithms and combinatorics; 17)
ISBN 3-540-64766-X

Mathematics Subject Classification (1991): 68-02, 68-Q, 68-R, 03-B99, 60-A99, 90-D99

ISSN 0937-5511
ISBN 3-540-64766-X Springer-Verlag Berlin Heidelberg New York

Typesetting: Typeset in LaTeX by the author. Reformatted by Kurt Mattes, Heidelberg, using a Springer TEX macro package
SPIN 10675255   46/3143 - 5 4 3 2 1 0 - Printed on acid-free paper

To Dana

Preface

> You can start by putting the DO NOT DISTURB sign.
>
> Cay, in *Desert Hearts* (1985).

The interplay between randomness and computation is one of the most fascinating scientific phenomena uncovered in the last couple of decades. This interplay is at the heart of modern cryptography and plays a fundamental role in complexity theory at large. Specifically, the interplay of randomness and computation is pivotal to several intriguing notions of probabilistic proof systems and is the focal of the computational approach to randomness. This book provides an introduction to these three, somewhat interwoven domains (i.e., cryptography, proofs and randomness).

Modern Cryptography. Whereas classical cryptography was confined to the art of designing and breaking encryption schemes (or "secrecy codes"), Modern Cryptography is concerned with the rigorous analysis of any system which should withstand malicious attempts to abuse it. We emphasize two aspects of the transition from classical to modern cryptography: (1) the widening of scope from one specific task to an utmost wide general class of tasks; and (2) the move from an engineering-art which strives on ad-hoc tricks to a scientific discipline based on rigorous approaches and techniques.

In this book we provide an introduction to the foundations of Modern Cryptography. We focus on the paradigms, approaches and techniques used to conceptualize, define and provide solutions to natural cryptographic problems. We also survey some of the fundamental results obtained using these paradigms, approaches and techniques. The emphasis of the exposition is on the need for and impact of a rigorous approach.

Probablistic Proof Systems. Various types of *probabilistic* proof systems have played a central role in the development of computer science in the last decade. These proof systems share a common (untraditional) feature – they carry a probability of error; yet, this probability is explicitly bounded and can be reduced by successive application of the proof system. The gain in allowing this untraditional relaxation is substantial, as demonstrated by three well known results regarding *interactive proofs*, *zero-knowledge proofs*, and *probabilistic checkable proofs*: In each of these cases, allowing a bounded probability of error makes the system much more powerful and useful than the traditional (errorless) counterparts.

Focusing on the three types of proof systems mentioned above, but going also beyond them, we survey the basic definitions and results regarding probabilistic proofs. Our exposition stresses both the similarities and differences between the various types of probabilistic proofs.

Pseudorandomness. A fresh view at the *question of randomness* was taken in the theory of computing: It has been postulated that a distribution is pseudorandom if it cannot be told apart from the uniform distribution by any efficient procedure. This paradigm, originally associating efficient procedures with polynomial-time algorithms, has been applied also with respect to a variety of limited classes of such distinguishing procedures.

Starting with the general paradigm, we survey the archetypical case of pseudorandom generators (withstanding any polynomial-time distinguisher), as well as generators withstanding space-bounded distinguishers, the derandomization of complexity classes such as $\mathcal{BPP}$, and some special-purpose generators.

An Underlying Assumption

Much of the contents of this book depends on the widely believed conjecture by which $\mathcal{P} \neq \mathcal{NP}$. This dependency is explicitly stated in some of the results which make even stronger assumptions (such as the existence of one-way functions), and is implicit in some results (such as the PCP Characterization of NP) which would become uninteresting if $\mathcal{P} = \mathcal{NP}$.

On the Nature of this Book

This book offers an introduction and extensive survey to each of the three areas mentioned above. It present both the basic notions and the most important (and sometimes advanced) results. The presentation is focused on the essentials and does not ellaborate on details. In some cases it offers a novel and illuminating perspective. The goal is to provide the reader with

1. A clear and structured overview of each of these areas.
2. Knowledge of the most important notions, ideas, techniques and results in each area.
3. Some new insights into each of these areas.

It is hoped that the book may be useful both to a beginner (who has only some background in the theory of computing), and to an expert in any of these areas.

Organization

In Chapter 1 we survey the basic concepts, definitions and results in cryptography. In particular, we survey the basic tools of cryptography – computational difficulty, pseudorandomness and zero-knowledge proofs – and the basic utilities – encryption, signatures, and general cryptographic protocols. Chapters 2 and 3 provides a wider perspective on two concepts mentioned in Chapter 1. Specifically, Chapter 2 surveys various types of probabilistic proof systems including interactive proofs, zero-knowledge proofs and probabilistically checkable proofs (PCP). (The overlap with Chapter 1 is small, and the presentation is quite different.) Likewise, Chapter 3 surveys various notions of pseudorandom generators, viewing the one discussed in Chapter 1 as an archetypical instantiation of a general paradigm.

The three chapters may be read independently of each other. In particular, each starts with an individual brief introduction to the respective subject matter. As hinted above, although the chapters do overlap, the perspectives taken in them are different. Specifically, Chapter 1 treats the theoretical foundations of a practical discipline, and so the presentation departs from practice and emphasizes *the importance of rigorous treatment for sound practice* (and not merely *per se*). In contrast, Chapters 2 and 3 depart from the theory of computing and emphasize the intellectual contents of the material (rather than its practical applicability). The fact that different perspectives co-exist in the same book, let alone in the same author, is indicative of the nature of the theory of computing.

The three chapters are augmented by four appendices and an extensive bibliography. Most importantly, Appendix A provides some basic background on computation and randomness.

We mention that important relations between randomness and computation were discovered also in other domains of the theory of computation. Some examples are given in Appendix B.

Appendix C provides proofs of two basic results; one being a folklore for which no proof has ever appeared, and the other for which the published proof is both too terse and more complex than the alternative presented here.

Acknowledgments

Much of the material was written while visiting the Laboratory for Computer Science of MIT.

A preliminary version of Chapter 1 has appeared in the proceedings of Advances in Cryptology – Crypto97, Springer's Lecture Notes in Computer Science (1997), Vol. 1294, pages 46–74.

Parts of the material presented in Chapter 2 have appeared in the proceedings of STACS97, Springer's Lecture Notes in Computer Science (1997), Vol. 1200, pages 595–611.

As for personal acknowledgments, I will only mention some of the people to whom I am most indebt for my professional development. These include Benny Chor, Shimon Even, Shafi Goldwasser, Leonid Levin, Silvio Micali, and Avi Wigderson.

> .... very little do we have and inclose which we can call our own in the deep sense of the word. We all have to accept and learn, either from our predecessors or from our contemporaries. Even the greatest genius would not have achieved much if he had wished to extract everything from inside himself. But there are many good people, who do not understand this, and spend half their lives wondering in darkness with their dreams of originality. I have known artists who were proud of not having followed any teacher and of owing everything only to their own genius. Such fools!
>
> [Goethe, *Conversations with Eckermann*, 17.2.1832]

Table of Contents

1. The Foundations of Modern Cryptography

It is possible to build a cabin with no foundations,
but not a lasting building.

Eng. Isidor Goldreich (1906–1995)

Summary − In our opinion, the Foundations of Cryptography are the paradigms, approaches and techniques used to conceptualize, define and provide solutions to natural cryptographic problems. In this chapter, we survey some of these paradigms, approaches and techniques as well as some of the fundamental results obtained using them. Special effort is made in attempt to dissolve common misconceptions regarding these paradigms and results.

Throughout this chapter, we assume some familiarity with modern cryptography; specifically with the basic settings of private-key and public-key cryptography. The unfamiliar reader is referred to Appendix A.4 for the adequate background.

1.1 Introduction

Cryptography is concerned with the construction of schemes which are robust against malicious attempts to make these schemes deviate from their prescribed functionality. Given a desired functionality, a cryptographer should design a scheme which not only satisfies the desired functionality under "normal operation", but also maintains this functionality in face of adversarial attempts which are devised after the cryptographer has completed his/her work. The fact that an adversary will devise its attack after the scheme has been specified, makes the design of such schemes very hard. In particular, the adversary will try to take actions other than the ones the designer had envisioned. Thus, our approach is that it makes little sense to make assumptions regarding the specific *strategy* that the adversary may use. The only assumptions which can be justified refer to the computational *abilities* of the adversary. Furthermore, it is our opinion that the design of cryptographic systems has to be based on *firm foundations*; whereas ad-hoc approaches and heuristics are a very dangerous way to go. A heuristic may make sense when the designer has a very good idea about the environment in which a scheme is to operate, yet a cryptographic scheme has to operate in a maliciously selected environment which typically transcends the designer's view.

Providing firm foundations to Cryptography has been a major research project in the last two decades. Indeed, the pioneering paper of Diffie and Hellman [121] should be considered the initiator of this project. Two major (interleaved) activities have been:

1. Definitional Activity: The identification, conceptualization and rigorous definition of cryptographic tasks which capture natural security concerns; and

2. Constructive Activity: The study and design of cryptographic schemes satisfying definitions as in (1).

The definitional activity provided a definition of secure encryption [198]. The reader may be surprised: *what is there to define* (beyond the basic setting formulated in [121])? Let us answer with a question (posed by Goldwasser and Micali [198]): *should an encryption scheme which leaks the first bit of the plaintext be considered secure?* Clearly, the answer is negative and so some naive conceptions regarding secure encryption (e.g., "a scheme is secure if it is infeasible to obtain the plaintext from the ciphertext when not given the decryption key") turn out to be unsatisfactory. The lesson is that even when a natural concern (e.g., "secure communication over insecure channels") has been identified, work still needs to be done towards a satisfactory (rigorous) definition of the underlying concept. The definitional activity also undertook the treatment of unforgeable signature schemes [200]: One result of the treatment was the refutation of a "folklore theorem" (attributed to Ron Rivest) by which "a signature scheme that is robust against chosen message attack cannot have a proof of security". The lesson here is that unclear/unsound formulations (i.e., those underlying the above folklore paradox) lead to false conclusions.

Another existing concept which was re-examined is the then-fuzzy notion of a "pseudorandom generator". Although ad-hoc "pseudorandom generators" which pass some ad-hoc statistical tests may be adequate for some statistical samplings, they are certainly inadequate for use in Cryptography: For example, sequences generated by linear congruential generators are easy to predict [74, 157] and endanger cryptographic applications even when not given in the clear [43]. The alternative suggested by Blum, Goldwasser, Micali and Yao [71, 198, 351] is a robust notion of pseudorandom generators – such a generator produces sequences which are *computationally indistinguishable* from truly random sequences, and thus, can replace truly random sequences in any practical application. We mention that the notion of computational indistinguishability has played a central role in the formulation of other cryptographic concepts (such as secure encryption and zero-knowledge).

The definitional activity has identified concepts which were not known before. One well-known example is the introduction of zero-knowledge proofs by Goldwasser, Micali and Rackoff [199]. A key paradigm crystallized in making the latter definition is the *simulation paradigm*: A party is said to have gained nothing from some extra information given to it if it can generate (i.e., simulate the receipt of) essentially the same information by itself (i.e., without being given this information). The simulation paradigm plays a central role in the related definitions of secure multi-party computations (with respect to varying settings such as in [268, 30, 197, 81, 50, 90]).

The definitional activity is an on-going process. Its more recent targets include session-key problems [50, 51, 34], mobile adversaries (a.k.a "Proactive Security") [293, 91, 215], Electronic Cash [96, 98, 156, 300, 328], Coercibility [87, 84], Threshold Cryptography [120], and more.

The constructive activity. As new definitions of cryptographic tasks emerged, the first challenge was to demonstrate that they can be achieved. Thus, the first goal of the constructive activity is to *demonstrate the plausibility* of obtaining certain goals. Standard assumptions such as that the RSA is hard to invert were used to construct secure public-key encryption schemes [198, 351] and unforgeable digital schemes [200]. We stress that assuming that RSA is hard to invert is different from assuming that RSA is a secure encryption scheme. Furthermore, plain RSA (alike any deterministic public-key encryption scheme) is not secure (as one can easily distinguish the encryption of one *predetermined* message from the encryption of another). Yet, RSA can be easily transformed into a secure public-key encryption scheme by using a construction [7] which is reminiscent of a common practice (of padding the message with random noise). We stress that the resulting scheme is not merely believed to be secure but rather its security is linked to a much simpler assumption (i.e., the assumption that RSA is hard to invert). Likewise, although plain RSA signing is vulnerable to "existential forgery" (and other attacks), RSA can be transformed into a signature scheme which is unforgeable (provided RSA is hard to invert) [200, 48]. Using the assumption that RSA is hard to invert, one can construct pseudorandom generators [71, 351], zero-knowledge proofs for any NP-statement [185], and multi-party protocols for securely computing any multi-variant function [353, 186].

A major misconception regarding theoretical work in Cryptography stems from not distinguishing work aimed at demonstrating the plausibility of obtaining certain goals from work aimed at suggesting paradigms and/or constructions which can be used in practice. For example, the general results concerning zero-knowledge proofs [185] and multi-party protocols [353, 186], mentioned above, are merely *claims of plausibility*: What they say is that any problem of the above type (i.e., any protocol problem as discussed in Section 1.7) can be solved in principle. This is a very valuable piece of information. Thus, if you have a specific problem which falls into the above category then you should know that the problem is solvable in principle. However, if you need to construct a real system then you should probably construct a solution from scratch (rather than employing the above general results). Typically, *some* tools developed towards solving the general problem may be useful in solving the specific problem. Thus, we distinguish three types of results:

1. *Plausibility results:* Here we refer to mere statements of the type "any NP-language has a zero-knowledge proof system" (cf., Goldreich, Micali and Wigderson [185]).
2. *Introduction of paradigms and techniques which may be applicable in practice:* Typical examples include construction paradigms as the "choose n out of $2n$ technique" of Rabin [302], the "authentication tree" of Merkle [261, 263], the "randomized encryption" paradigm of Goldwasser and Micali [198], proof techniques as the "hybrid argument" of [198] (cf., [170, Sec. 3.2.3]), and many others.
3. *Presentation of schemes which are suitable for practical applications:* Typical examples include the public-key encryption schemes of Blum and Goldwasser [68], the digital signature schemes of [131, 127, 110], the session-key protocols of [50, 51], and many others.

Typically, it is quite easy to determine to which of the above categories a specific technical contribution belongs. Unfortunately, the classification is not always stated in the paper; however, it is typically evident from the construction. We stress that all results we are aware of (and in particular all results cited in this chapter), come with an explicit construction. Furthermore, the security of the resulting construction is explicitly related to the complexity of certain intractable tasks. In contrast to some uninformed beliefs, for each of these results there is an explicit translation of concrete intractability assumptions (on which the scheme is based) into lower bounds on the amount of work required to violate the security of the resulting scheme.[1] We stress that this translation can be invoked for any value of the security parameter. Doing so determines whether a specific construction is adequate for a specific application under specific reasonable intractability assumptions. In many cases the answer is in the affirmative, but in general this does depend on the specific construction as well as on the specific value of the security parameter and on what is reasonable to assume for this value. When we say that a result is suitable for practical applications (i.e., belongs to Type 3 above), we mean that it offers reasonable security for reasonable implementation values of the security parameter and reasonable assumptions.

Other activities. This chapter is focused on the definitional and constructive activities mentioned above. Other activities in the foundations of cryp-

---

[1] The only exception to the latter statement is Levin's observation regarding the existence of a *universal one-way function* (cf., [245] and [170, Sec. 2.4.1]).

tography include the exploration of new directions and the marking of limitations. For example, we mention novel modes of operation such as split-entities [58, 120, 265], batching operations [149], off-line/on-line signing [131] and Incremental Cryptography [38, 39]. On the limitation side, we mention [220, 179]. In particular, [220] indicates that certain tasks (e.g., secret key exchange) are unlikely to be achieved by using a one-way function in a "black-box manner".

Organization

Although encryption, signatures and secure protocols are the primary tasks of Cryptography, we start our presentation with basic paradigms and tools such as computational difficulty (Section 1.2), pseudorandomness (Section 1.3) and zero-knowledge (Section 1.4). Once these are presented, we turn to encryption (Section 1.5), signatures (Section 1.6) and secure protocols (Section 1.7). We conclude with some notes (Section 1.8), a short historical perspective (Section 1.9), two suggestions for future research (Section 1.10) and some suggestions for further reading (Section 1.11).

1.2 Central Paradigms

Modern Cryptography, as surveyed here, is concerned with the construction of *efficient* schemes for which it is *infeasible* to violate the security feature. Thus, we need a notion of efficient computations as well as a notion of infeasible ones. The computations of the legitimate users of the scheme ought be efficient; whereas violating the security features (via an adversary) ought to be infeasible. Our notions of efficient and infeasible computations are "asymptotic": They refer to the running time as a function of the security parameter. This is done in order to avoid cumbersome formulations which refer to the actual running-time on a specific model for specific values of the security parameter. As discussed above one can easily derive such specific statements from the asymptotic treatment. Actually, the term "asymptotic" is misleading since, from the functional treatment of the running-time (as a function of the security parameter), one can derive statements for ANY value of the security parameter.

Efficient computations are commonly modeled by computations which are polynomial-time in the security parameter. The polynomial bounding the running-time of the legitimate user's strategy is fixed and typically explicit and small (still in some cases it is indeed a valuable goal to make it even smaller). Here (i.e., when referring to the complexity of the legitimate user) we are in the same situation as in any algorithmic research. Things are different when referring to our assumptions regarding the computational resources of the adversary. A common approach is to postulate that the latter are polynomial-time too, where the polynomial is NOT a-priori specified. In other words, the adversary is restricted to the class of efficient computations and anything beyond this is considered to be infeasible. Although many definitions explicitly refer to this convention, this convention is INESSENTIAL to any of the results known in the area. In all cases, a more general (and yet more cumbersome) statement can be made by referring to adversaries of running-time bounded by any function (or class of functions). For example, for any function $T : \mathbb{N} \mapsto \mathbb{N}$ (e.g., $T(n) = 2^{\sqrt[3]{n}}$), we may consider adversaries which on security parameter n run for at most $T(n)$ steps. Doing so we (implicitly) define as infeasible any computation which (on security parameter n) requires more than $T(n)$ steps. A typical result has the form[2]

> If RSA with n-bit moduli cannot be inverted in time $T(n)$ then the following construction (using security parameter n) is secure against adversaries operating in time $T'(n) = T(g(n))/f(n)$, where f and g^{-1} are explicitly given polynomials.

However, most papers prefer to present a simplified statement of the form "if RSA cannot be inverted in polynomial-time then the following construction is secure against polynomial-time adversaries". This is unfortunate since it is the specific functions f and g, which are (sometimes explicit and) always implicit in the proof, that determine the practicality of the construction.[3] The smaller f and g^{-1}, the better. Our rule of thumb is that results with $g^{-1}(n) = O(n)$ (e.g., $g(n) = n/2$) are practical, whereas results with, say, $g^{-1}(n) = n^4$ (i.e., $g(n) = \sqrt[4]{n}$) are to be considered merely plausibility results.

Lastly, we consider the notion of a negligible probability. The idea behind this notion is to have a robust notion of rareness: A rare event should occur rarely even if we repeat the experiment for a feasible number of times. That is, if we consider any polynomial-time computation to be feasible then any function $f : \mathbb{N} \mapsto \mathbb{N}$ so that $(1 - f(n))^{p(n)} > 0.99$, for any polynomial p, is considered negligible (i.e., f is negligible if for any polynomial p the function $f(\cdot)$ is bounded above by $1/p(\cdot)$). However, if we consider the function $T(n)$ to provide our notion of infeasible computation then functions bounded above by $1/T(n)$ are considered negligible (in n).

---

[2] Actually, the form below is over-simplified. The actual statement refers also to the success probabilities of both attacks. It reads: If RSA with n-bit moduli cannot be inverted in time $T(n)$, with success probability greater than $\epsilon(n)$, then the following construction (using security parameter n) cannot be broken by adversaries operating in time $T'(n)$ with success probability greater than $\epsilon'(n)$, where $T'(n)$ and $\epsilon'(n)$ are related to $T(g(n))$ and $\epsilon(g(n))$ via explicit polynomial expressions and g^{-1} is an explicitly given polynomial. Specifically, $T(g(n)) = \text{poly}(n, T'(n))/\text{poly}(\epsilon'(n))$ and $\epsilon(g(n)) = \text{poly}(\epsilon'(n))/\text{poly}(n, T'(n))$. Typically, $T(g(n)) = \text{poly}(n/\epsilon'(n)) \cdot T'(n)$ and $\epsilon(g(n)) = \text{poly}(\epsilon'(n))/\text{poly}(T'(n))$.

[3] The importance of *explicitly* relating the security of the resulting scheme to the quantified intractability assumption has been advocated (and practiced) in a sequence of recent works by Bellare and Rogaway (cf., [47, p. 343]).

In the rest of this chapter we adopt the simpler convention of defining infeasible computations as ones which cannot be conducted in polynomial-time. (However, we explicitly state the level of practicality of each of the results presented.) The interested reader is referred to [251] for a more general treatment.

1.2.1 Computational Difficulty

Modern Cryptography is concerned with the construction of schemes which are easy to operate (properly) but hard to foil. Thus, a complexity gap (i.e., between the complexity of proper usage and the complexity of defeating the prescribed functionality) lies in the heart of Modern Cryptography. However, gaps as required for Modern Cryptography are not known to exist – they are only widely believed to exist. Indeed, almost all of Modern Cryptography rises or falls with the question of whether one-way functions exist (e.g., see [211, 174, 313, 272, 185] for positive results and [245, 313, 292] for negative ones). One-way functions are functions which are easy to evaluate but hard (on the average) to invert.

Definition 1.1 (one-way functions [121]): *A function $f : \{0,1\}^* \mapsto \{0,1\}^*$ is called* one-way *if*

- *easy direction: there is an efficient algorithm which on input x outputs $f(x)$.*
- *hard direction: given $f(x)$, where x is uniformly selected, it is infeasible to find, with non-negligible probability, a preimage of $f(x)$. That is, any feasible algorithm which tries to invert f may succeed only with negligible probability, where the probability is taken over the choices of x and the algorithm's coin tosses.*

Warning: The above definition, as well as all other definitions in this chapter, avoids some technicalities and so is imprecise.[4] For precise definitions, the interested reader is referred to other texts (see Section 1.11).

Some known constructions require special types of one-way functions: One-way permutations are length-preserving 1-1 (one-way) functions, commonly viewed as collections of finite permutations, each having its own domain. Such a collection is specified by efficient algorithms for selecting a (succinct representation of a) permutation from the collection, for sampling the domain of a given permutation, and – of course – for evaluating a given permutation. Some construction require one-way permutations with a *trapdoor*. That is, the permutation-selection algorithm generates a representation of the permutation along with some trapdoor information, so that it is easy

---

[4] In this case, the missing technicality is requiring that f does not shrink its input too much; that is, $|x| = \mathrm{poly}(|f(x)|)$, $\forall x$. (Otherwise, the inversion task is infeasible for trivial reasons, and has no useful consequences.)

to invert the permutation given this trapdoor but it remains hard to do so when only given the representation of the permutation. The RSA (cf., [312] or Appendix A.4) is a popular candidate trapdoor permutation.

1.2.2 Computational Indistinguishability

A central notion in Modern Cryptography is that of "effective similarity". The underlying idea is that we do not care if objects are equal or not – all we care is whether a difference between the objects can be observed by a feasible computation. In case the answer is negative, we may say that the two objects are equivalent as far as any practical application is concerned. Indeed, it will be our common practice to interchange such (computationally indistinguishable) objects.

Definition 1.2 (computational indistinguishability [198, 351]): *Let $X = \{X_n\}_{n\in\mathbb{N}}$ and $Y = \{Y_n\}_{n\in\mathbb{N}}$ be probability ensembles such that each X_n and Y_n ranges over strings of length n. We say that X and Y are* computationally indistinguishable *if for every feasible algorithm A the difference*

$$d_A(n) \stackrel{\text{def}}{=} |\Pr[A(X_n)=1] - \Pr[A(Y_n)=1]|$$

is a negligible function in n.

1.2.3 The Simulation Paradigm

A key question regarding the modeling of security concerns is how to express the intuitive requirement that an adversary "gains nothing substantial" by deviating from the prescribed behavior of an honest user. The approach initiated in [198, 199] is that the adversary *gains nothing* if whatever it can obtain by deviating from the prescribed honest behavior can also be obtained in an appropriately defined "ideal model". The definition of the "ideal model" captures what we want to achieve in terms of security, and so is specific to the security concern to be addressed. For example, an encryption scheme is considered secure (against eavesdropping) if an adversary which eavesdrops a channel on which encrypted messages are sent, gains nothing over a user which does not tap this channel. Thus, the encryption scheme "simulates" an ideal private channel between parties.

A notable property of the above simulation paradigm, as well as of the entire approach surveyed here, is that this approach is very liberal with respect to its view of the abilities of the adversary as well as to what might constitute a gain for the adversary. For example, we consider an encryption scheme to be secure only if it can simulate a private channel. Indeed, failure to provide such a simulation does NOT necessarily mean that the encryption scheme can be "broken" in some intuitively harmful sense. Thus, it seems that our approach to defining security is overly cautious. However, it seems impossible to come up with definitions of security which distinguish "breaking the scheme in a harmful sense" from "breaking it in a non-harmful sense". Firstly, even in a specific application, the notion of a "harmful breaking" is a very evasive one (and typically becomes clear only after the system is broken). More importantly, whatever is harmful is application-dependent, whereas a good definition of security ought to be application independent (as otherwise using the scheme in any new application will require a full re-evaluation of its security).

1.3 Pseudorandomness

In practice "pseudorandom" sequences are used instead of truly random sequences in many applications. The underlying belief is that if an (efficient) application performs well when using a truly random sequence then it will perform essentially as well when using a "pseudorandom" sequence. However, this belief is not supported by previous characterizations of "pseudorandomness" (e.g., such as passing the statistical tests in Knuth's book [234] or having large linear-complexity[5]). In contrast, the above belief is an easy corollary of defining pseudorandom distributions as ones which are computationally indistinguishable from uniform distributions.

1.3.1 The Basics

We are interested in pseudorandom sequences which can be determined by and generated from short random seeds. That is,

Definition 1.3 (pseudorandom generator [71, 351]): *Let $\ell : \mathbb{N} \mapsto \mathbb{N}$ be so that $\ell(n) > n$, $\forall n$. A **pseudorandom generator**, with **stretch** function ℓ, is an efficient* (deterministic) *algorithm which on input a random n-bit seed outputs a $\ell(n)$-bit sequence which is computationally indistinguishable from a uniformly chosen $\ell(n)$-bit sequence.*

We stress that pseudorandom sequences can replace truly random sequences not only in "ordinary" computations but also in cryptographic ones. That is, ANY cryptographic application which is secure when the legitimate parties use truly random sequences, is also secure when the legitimate parties use pseudorandom sequences. Various cryptographic applications of pseudorandom generators will be presented in the sequel, but first let us consider the construction of pseudorandom generators. A key paradigm is presented next. It uses the notion of a *hard-core* predicate [71] of a (one-way) function: The predicate b is a **hard-core** of the function f if b is easy to evaluate but $b(x)$

---

[5] The linear complexity of a sequence is defined as the length of the shortest Linear Feedback Shift Register which produces it. See [204].

is hard to predict from $f(x)$. That is, it is infeasible, given $f(x)$ when x is uniformly chosen, to predict $b(x)$ substantially better than with probability $1/2$. Intuitively, b "inherits *in a concentrated sense*" the difficulty of inverting f. (Note that if b is a hard-core of an efficiently computable 1-1 function f then f must be one-way.)

The iteration paradigm [71]: Let f be a 1-1 function which is length-preserving and efficiently computable, and b be a hard-core predicate of f. Then

$$G(s) = b(s) \cdot b(f(s)) \cdots b(f^{\ell(|s|)-1}(s))$$

is a pseudorandom generator (with stretch function ℓ), where $f^{i+1}(x) \overset{\text{def}}{=} f(f^i(x))$ and $f^0(x) \overset{\text{def}}{=} x$. As a concrete example, consider the permutation $x \mapsto x^2 \bmod N$, where N is the product of two primes each congruent to 3 (mod 4). We have $G_N(s) = \mathrm{lsb}(s) \cdot \mathrm{lsb}(s^2 \bmod N) \cdots \mathrm{lsb}(s^{2^{\ell(|s|)-1}} \bmod N)$, where $\mathrm{lsb}(x)$ is the least significant bit of x (which by [7, 346] is a hard-core of the modular squaring function). We note that for any one-way permutation f', the inner-product mod 2 of x and r is a hard-core of $f(x,r) = (f'(x),r)$ [182]. Thus, using any one-way permutation, we can easily construct pseudorandom generators.

The iteration paradigm is even more beneficial when one has a hard-core function rather than a hard-core predicate: h is called a *hard-core function* of f if h is easy to evaluate but, for a random $x \in \{0,1\}^*$, the distribution $f(x) \cdot h(x)$ is pseudorandom. (Note that a hard-core predicate is a special case.) Using a hard-core function h for f, we obtain the pseudorandom generator $G'(s) = h(s) \cdot h(f(s)) \cdot h(f^2(s)) \cdots$. In particular, assuming the intractability of the subset sum problem (for suitable densities) or of the decoding of random linear codes, this paradigm was used in [219, 151] to construct very efficient pseudorandom generators. Alternatively, encouraged by the results in [7, 213], we conjecture that the first $n/2$ least significant bits of the argument constitute a hard-core function of the modular squaring function for n-bit long moduli. This conjecture yields an efficient pseudorandom generator: $G'_N(s) = \mathrm{LSB}_N(s) \cdot \mathrm{LSB}_N(s^2 \bmod N) \cdot \mathrm{LSB}_N(s^4 \bmod N) \cdots$, where $\mathrm{LSB}_N(x)$ denotes the $0.5 \log_2 N$ least significant bits of x.

A plausibility result [211]: Pseudorandom generators exist if (and only if) one-way functions exist. Unlike the construction of pseudorandom generators from one-way permutations, the known construction of pseudorandom generators from *arbitrary* one-way functions has no practical significance. It is indeed an important open problem to provide an alternative construction which may be practical and still utilize an *arbitrary* one-way function.

1.3.2 Pseudorandom Functions

Pseudorandom generators allow to efficiently generate long pseudorandom sequences from short random seeds. Pseudorandom functions (defined below) are even more powerful: They allow efficient direct access to a huge pseudorandom sequence (which is not feasible to scan bit-by-bit). Put in other words, pseudorandom functions can replace truly random functions in any application where the function is used in a black-box fashion (i.e., the adversary may indirectly obtain the value of the function at arguments of its choice, but does not have the description of the function and so is not able to evaluate the function by itself).[6]

Definition 1.4 (pseudorandom functions [174]): *A* pseudorandom function *is an efficient* (deterministic) *algorithm which given an n-bit seed, s, and an n-bit argument, x, returns an n-bit string, denoted $f_s(x)$, so that it is infeasible to distinguish the responses of f_s, for a uniformly chosen s, from the responses of a truly random function.*

That is, the distinguisher is given access to a function and is required to distinguish a random function $f : \{0,1\}^n \mapsto \{0,1\}^n$ from a function chosen uniformly in $\{f_s : s \in \{0,1\}^n\}$. We stress that in the latter case the distinguisher is NOT given the description of the function f_s (i.e., the seed s), but rather may obtain the value of f_s on any n-bit string of its choice.[7]

Pseudorandom functions are a very useful cryptographic tool (cf., [175, 165] and Section 1.5): One may first design a cryptographic scheme assuming that the legitimate users have black-box access to a random function, and next implement the random function using a pseudorandom function. We stress that the description of the pseudorandom function is given to the legitimate users but not to the adversary. (The adversary may be able to obtain from the legitimate users the value of the function on arguments of its choice, but not the function's description.)

From pseudorandom generators to pseudorandom functions [174]:
Let G be a pseudorandom generator with stretching function $\ell(n) = 2n$, and let $G_0(s)$ (resp., $G_1(s)$) denote the first (resp., last) n bits in $G(s)$ where $s \in \{0,1\}^n$. We define the function ensemble $\{f_s : \{0,1\}^{|s|} \mapsto \{0,1\}^{|s|}\}$, where $f_s(\sigma_{|s|} \cdots \sigma_2 \sigma_1) = G_{\sigma_{|s|}}(\cdots G_{\sigma_2}(G_{\sigma_1}(s)) \cdots)$. This ensemble is pseudorandom.

Alternative constructions of pseudorandom functions have been suggested in [276, 278].

---

[6] This is different from the *Random Oracle Model* of [49], where the adversary has a *direct* access to a random oracle (that is later "implemented" by a function, the description of which is given also to the adversary).

[7] Typically, the distinguisher stands for an adversary that attacks a system which uses a pseudorandom function. The values of the function on arguments of the adversary's choice are obtained from the legitimate users of the system who, *unlike the adversary*, know the seed s. The definition implies that the adversary will not be more successful in its attack than it could have been if the system was to use a truly random function. Needless to say that the latter system is merely a *Gedanken Experiment* (it cannot be implemented since it is infeasible to even store a truly random function).

1.4 Zero-Knowledge

Loosely speaking, zero-knowledge proofs are proofs which yield nothing beyond the validity of the assertion. That is, a verifier obtaining such a proof only gains conviction in the validity of the assertion. Using the simulation paradigm this requirement is stated by postulating that anything that is feasibly computable from a zero-knowledge proof is also feasibly computable from the valid assertion alone.

1.4.1 The Basics

The above informal paragraph refers to proofs as to interactive and randomized processes.[8] That is, here a proof is a (multi-round) protocol for two parties, called verifier and prover, in which the prover wishes to convince the verifier of the validity of a given assertion. Such an *interactive proof* should allow the prover to convince the verifier of the validity of any true assertion, whereas NO prover strategy may fool the verifier to accept false assertions. Both the above *completeness* and *soundness* conditions should hold with high probability (i.e., a negligible error probability is allowed). The prescribed verifier strategy is required to be efficient. No such requirement is made with respect to the prover strategy; yet we will be interested in "relatively efficient" prover strategies (see below). Zero-knowledge is a property of some prover-strategies. More generally, we consider interactive machines which yield no knowledge while interacting with an arbitrary feasible adversary on a common input taken from a predetermined set (in our case the set of valid assertions).

Definition 1.5 (zero-knowledge [199]): *A strategy A is* zero-knowledge *on inputs from S if, for every feasible strategy B^*, there exists a feasible computation C^* so that the following two probability ensembles are computationally indistinguishable:*

1. $\{(A, B^*)(x)\}_{x \in S} \stackrel{\text{def}}{=}$ *the output of B^* when interacting with A on common input $x \in S$; and*
2. $\{C^*(x)\}_{x \in S} \stackrel{\text{def}}{=}$ *the output of C^* on input $x \in S$.*

Note that whereas A and B^* above are interactive strategies, C^* is a non-interactive computation. The above definition does NOT account for auxiliary information which an adversary may have prior to entering the interaction. Accounting for such auxiliary information is essential for using zero-knowledge proofs as subprotocols inside larger protocols (see [179, 187]). Another concern is that we prefer that the complexity of C^* be bounded

---

[8] The formulation applies also to "proofs" in the ordinary sense of being strings (i.e., NP-proofs). However, zero-knowledge NP-proofs exist only in a trivial manner (i.e., for languages in $\mathcal{BPP}$).

as a function of the complexity of B^*. Both concerns are taken care of by a more strict notion of zero-knowledge presented next.

Definition 1.6 (zero-knowledge, revisited [187]): *A strategy A is* black-box zero-knowledge *on inputs from S if there exists an efficient* (universal) *subroutine-calling algorithm U so that for every feasible strategy B^*, the probability ensembles $\{(A, B^*)(x)\}_{x \in S}$ and $\{U^{B^*}(x)\}_{x \in S}$ are computationally indistinguishable, where U^{B^*} is algorithm U using strategy B^* as a subroutine.*

Note that the running time of U^{B^*} is at most the running-time of U times the running-time of B^*. Actually, the first term may be replaced by the number of times U invokes the subroutine. Almost all known zero-knowledge proofs are in fact black-box zero-knowledge.[9]

A general plausibility result [185]: Assuming the existence of commitment schemes, there exist (black-box) zero-knowledge proofs for membership in any NP-language. Furthermore, the prescribed prover strategy is efficient, provided it is given an NP-witness to the assertion to be proven. This makes zero-knowledge a very powerful tool in the design of cryptographic schemes and protocols.

Zero-knowledge as a tool: In a typical cryptographic setting, a user, referred to as A, has a secret and is supposed to take some steps depending on its secret. The question is how can other users verify that A indeed took the correct steps (as determined by A's secret and some publicly known information). Indeed, if A discloses its secret then anybody can verify that it took the correct steps. However, A does not want to reveal its secret. Using zero-knowledge proofs we can satisfy both conflicting requirements. That is, A can prove in zero-knowledge that it took the correct steps. Note that A's claim to having taken the correct steps is an NP-assertion and that A has an NP-witness to its validity (i.e., its secret!). Thus, by the above result, it is possible for A to efficiently prove the correctness of its actions without yielding anything about its secret. (However, in practice one may want to design a specific zero-knowledge proof, tailored to the specific application and so being more efficient, rather than invoking the general result above. Thus, the development of techniques for the construction of efficient zero-knowledge proof systems is still of interest – see, for example, [178, 79, 144, 232, 117, 111, 290, 316, 192, 112].)

1.4.2 Some Variants

Perfect zero-knowledge arguments: This term captures two deviations from the above definition; the first being a strengthening and the second

---

[9] The only exception we are aware of are contrived protocols constructed in [179] (for the purpose of separating Definitions 1.5 and 1.6), and the 3-message protocol of [206] (designed – using non-standard assumptions – so to bypass the "triviality result" of [179] regarding 3-message *black-box* zero-knowledge proofs).

being a weakening. Perfect zero-knowledge strategies are such for which the ensembles in Definition 1.5 are identically distributed (rather than computationally indistinguishable). This means that the zero-knowledge clause holds regardless of the computational abilities of the adversary. However, *arguments* (aka *computationally sound proofs*) differ from interactive proofs in having a weaker soundness clause: it is infeasible (rather than impossible) to fool the verifier to accept false assertion (except with negligible probability) [77]. Perfect zero-knowledge arguments for NP were constructed using any one-way permutation [275].

Non-Interactive zero-knowledge proofs [66, 143]: Here the interaction between the prover and the verifier consists of the prover sending a single message to the verifier (as in "classical proofs"). In addition, both players have access to a "random reference string" which is postulated to be uniformly selected. Non-interactive zero-knowledge proofs are useful in applications where one of the parties may be trusted to select the abovementioned reference string (e.g., see Section 1.5.3). Non-interactive zero-knowledge arguments for NP were constructed using any trapdoor permutation [143, 233].

Zero-knowledge proofs of knowledge [199, 150, 36]: Loosely speaking, a system for proofs of knowledge guarantees that whenever the verifier is convinced that the prover knows X, this X can be efficiently extracted from the prover's strategy. One natural application of (zero-knowledge) proofs of knowledge is for *identification* [150, 137]. Figure 1.1 depicts the Fiat-Shamir Identification Scheme [150] (which is based on the Goldwasser-Micali-Rackoff zero-knowledge proof system for Quadratic Residuosity [199]).

---

universal parameter: A composite N, product of two (secret) primes.
private-key (of user U): A uniformly chosen $x_u \in \{1, \ldots, N\}$.
public-key: (of user U): $y_u = x_u^2 \bmod N$.
protocol for user U to identify itself.   (basic version)

    1. Prover uniformly select $r \in \{1, \ldots, N\}$, and sends $s \stackrel{\text{def}}{=} r^2 \bmod N$ to the verifier.
    2. The verifier uniformly select a challenge $\sigma \in \{0,1\}$, and sends it to the prover.
    3. Prover replies with $z \stackrel{\text{def}}{=} r \cdot x_u^\sigma \bmod N$.
    4. The verifier accepts if and only if $z^2 \equiv s \cdot y_u^\sigma \pmod{N}$.

The above protocol is a zero-knowledge proof of knowledge of a modular square root of y_u. Since U is supposedly the only party knowing the square root of y_u, succeeding in this protocol is taken as evidence that the prover is U. The zero-knowledge clause guarantees that interacting with U according to the protocol, does not leak knowledge which may be used to impersonate U. For more details see [199, 150, 137].

Fig. 1.1. The Fiat–Shamir Identification Scheme [150] – basic version.

Relaxations of Zero-knowledge: Important relaxations of zero-knowledge were presented in [145]. Specifically, in *witness indistinguishable* proofs it is infeasible to tell which NP-witness to the assertion the prover is using. Unlike zero-knowledge proofs, this notion is closed under parallel composition. Furthermore, this relaxation suffices for some applications in which one may originally think of using zero-knowledge proofs.

1.5 Encryption

Both Private-Key and Public-Key encryption schemes consists of three efficient algorithms: *key generation, encryption* and *decryption*. The difference between the two types is reflected in the definition of security – the security of a public-key encryption scheme should hold also when the adversary is given the encryption key, whereas this is not required for private-key encryption scheme. Thus, public-key encryption schemes allow each user to broadcast its encryption key so that any user may send it encrypted messages (without needing to first agree on a private encryption-key with the receiver). Below we present definitions of security for private-key encryption schemes. The public-key analogies can be easily derived by considering adversaries which get the encryption key as additional input. (For private-key encryption schemes we may assume, without loss of generality, that the encryption key is identical to the decryption key.)

1.5.1 Definitions

For simplicity we consider only the encryption of a single message; however this message may be longer than the key (which rules out information-theoretic secrecy [323]). We present two equivalent definitions of security. The first, called *semantic security*, is a computational analogue of Shannon's definition of *perfect secrecy* [323]. The second definition views secure encryption schemes as ones for which it is infeasible to distinguish encryptions of any (known) pair of messages (e.g., the all-zeros message and the all-ones message). The latter definition is technical in nature and is referred to as *indistinguishability of encryptions*.

We stress that the definitions presented below go way beyond saying that it is infeasible to recover the plaintext from the ciphertext. The latter statement is indeed a minimal requirement from a secure encryption scheme, but we claim that it is way too weak a requirement: An encryption scheme is typically used in applications where obtaining specific partial information on the plaintext endangers the security of the application. When designing an application-independent encryption scheme, we do not know which partial information endangers the application and which does not. Furthermore, even if one wants to design an encryption scheme tailored to one's own specific applications, it is rare (to say the least) that one has a precise characterization of all possible partial information which endanger these applications. Thus, we require that it is infeasible to obtain any information about the plaintext from the ciphertext. Furthermore, in most applications the plaintext may not be uniformly distributed and some a-priori information regarding it is available to the adversary. We require that the secrecy of all partial information is preserved also in such a case. That is, even in presence of a-priori information on the plaintext, it is infeasible to obtain any (new) information about the plaintext from the ciphertext (beyond what is feasible to obtain from the a-priori information on the plaintext). The definition of semantic security postulates all of this. The equivalent definition of indistinguishability of encryptions is useful in demonstrating the security of candidate constructions as well as for arguing about their usage as part of larger protocols.

The actual definitions: In both definitions we consider (feasible) adversaries which obtain, in addition to the ciphertext, also auxiliary information which may depend on the potential plaintexts (but not on the key). By $E(x)$ we denote the distribution of encryptions of x, when the key is selected at random. To simplify the exposition, let us assume that on security parameter n the key generation produces a key of length n, whereas the scheme is used to encrypt messages of length n^2.

Definition 1.7 (semantic security (following [198])): *An encryption scheme is* semantically secure *if for every feasible algorithm, A, there exists a feasible algorithm B so that for every two functions $f, h : \{0,1\}^* \mapsto \{0,1\}^*$ and all probability ensembles, $\{X_n\}_{n \in \mathbb{N}}$, where X_n ranges over $\{0,1\}^{n^2}$,*

$$\Pr[A(E(X_n), h(X_n)) = f(X_n)] < \Pr[B(h(X_n)) = f(X_n)] + \mu(n)$$

where μ is a negligible function. Furthermore, the complexity of B should be related to that of A.

What this definition says is that a feasible adversary does not gain anything by looking at the ciphertext. That is, whatever information (captured by the function f) it tries to compute from the ciphertext, can be essentially computed as efficiently from the available a-priori information (captured by the function h). In particular, the ciphertext does not help in (feasibly) computing the least significant bit of the plaintext or any other information regarding the plaintext. This holds for any distribution of plaintexts (captured by the random variable X_n).

Definition 1.8 (indistinguishability of encryptions (following [198])): *An encryption scheme has* indistinguishable encryptions *if for every feasible algorithm, A, and all sequences of triples, $(x_n, y_n, z_n)_{n \in \mathbb{N}}$, where $|x_n| = |y_n| = n^2$ and $|z_n|$ is of feasible (in n) length, the difference*

$$d_A(n) \stackrel{\text{def}}{=} |\Pr[A(E(x_n), z_n) = 1] - \Pr[A(E(y_n), z_n) = 1]|$$

is a negligible function in n.

In particular, z_n may equal (x_n, y_n). Thus, it is infeasible to distinguish the encryptions of any two fix messages such as the all-zero message and the all-ones message.

Probabilistic Encryption: It is easy to see that a secure *public-key* encryption scheme must employ a probabilistic (i.e., randomized) encryption algorithm. Otherwise, given the encryption key as (additional) input, it is easy to distinguish the encryption of the all-zero message from the encryption of the all-ones message. The same holds for *private-key* encryption schemes when considering the security of encrypting several messages (rather than a single message as done above).[10] This explains the linkage between the above robust security definitions and the *randomization paradigm* (discussed below).

1.5.2 Constructions

It is common practice to use "pseudorandom generators" as a basis for private-key stream ciphers. We stress that this is a very dangerous practice when the "pseudorandom generator" is easy to predict (such as the linear congruential generator or some modifications of it which output a constant fraction of the bits of each resulting number – see [74, 157]). However, this common practice becomes sound provided one uses pseudorandom generators (as defined in Section 1.3). An alternative, more flexible construction follows.

Private-Key Encryption based on Pseudorandom Functions: The key generation algorithm consists of selecting a seed, denoted s, for such a function, denoted f_s. To encrypt a message $x \in \{0, 1\}^n$ (using key s), the encryption algorithm uniformly selects a string $r \in \{0, 1\}^n$ and produces the ciphertext $(r, x \oplus f_s(r))$. To decrypt the ciphertext (r, y) (using key s), the decryption algorithm just computes $y \oplus f_s(r)$. The proof of security of this encryption scheme consists of two steps (suggested as a general methodology in Section 1.3):

1. Prove that an idealized version of the scheme, in which one uses a uniformly selected function $f: \{0, 1\}^n \mapsto \{0, 1\}^n$, rather than the pseudorandom function f_s, is secure.
2. Conclude that the real scheme (as presented above) is secure (since otherwise one could distinguish a pseudorandom function from a truly random one).

Note that we could have gotten rid of the randomization if we had allowed the encryption algorithm to be history dependent (e.g., use a counter in the role of r). Furthermore, if the encryption scheme is used for FIFO communication between the parties and both can maintain the counter value then there is no need for the sender to send the counter value.

---

[10] Here, for example, using a deterministic encryption algorithm allows the adversary to distinguish two encryptions of the same message from the encryptions of a pair of different messages.

The randomization paradigm [198]: We demonstrate this paradigm by presenting several constructions of public-key encryption schemes. First, suppose we have a trapdoor one-way permutation, $\{p_\alpha\}_\alpha$, and a hard-core predicate, b, for it.[11] The key generation algorithm consists of selecting at random a permutation p_α together with a trapdoor for it: The permutation (or rather its description) serves as the public-key, whereas the trapdoor serves as the private-key. To encrypt a single bit σ (using public key p_α), the encryption algorithm uniformly selects an element, r, in the domain of p_α and produces the ciphertext $(p_\alpha(r), \sigma \oplus b(r))$. To decrypt the ciphertext (y, τ) (using the private key), the decryption algorithm just computes $\tau \oplus b(p_\alpha^{-1}(y))$ (where the inverse is computed using the trapdoor (i.e., private-key)). The above scheme is quite wasteful in bandwidth; however, the paradigm underlying its construction is valuable in practice. Following are two important examples.

First, we note that it is better to randomly pad messages (say using padding equal in length to the message) before encrypting them using RSA, than to employ RSA on the plain message. Such a heuristic could be placed on firm grounds if a conjecture analogous to the one mentioned in Section 1.3 is supported. That is, assume that the first $n/2$ least significant bits of the argument constitute a hard-core function of RSA with n-bit long moduli. Then, encrypting $n/2$-bit messages by padding the message with $n/2$ random bits and applying RSA (with an n-bit moduli) on the result constitutes a secure public-key encryption system, hereafter referred to as Randomized RSA.

Secondly, following [68], we present an alternative public-key encryption scheme, which can be based on any trapdoor permutation. In contrast to the scheme presented above, the current scheme is not wasteful in bandwidth. The encryption scheme augments the construction of a pseudorandom generator, given in Section 1.3, as follows. The key-generation algorithm consists of selecting at random a permutation p_α together with a trapdoor. To encrypt the n-bit string x (using public key p_α), the encryption algorithm uniformly selects an element, s, in the domain of p_α and produces the ciphertext $(p_\alpha^n(s), x \oplus G_\alpha(s))$, where $G_\alpha(s) = b(s) \cdot b(p_\alpha(s)) \cdots b(p_\alpha^{n-1}(s))$. (We use the notation $p_\alpha^{i+1}(x) = p_\alpha(p_\alpha^i(x))$ and $p_\alpha^{-(i+1)}(x) = p_\alpha^{-1}(p_\alpha^{-i}(x))$.) To decrypt the ciphertext (y, z) (using the private key), the decryption algorithm first recovers $s = p_\alpha^{-n}(y)$ and then outputs $z \oplus G_\alpha(s)$.

Assuming that factoring Blum Integers (i.e., products of two primes each congruent to 3 (mod 4)) is hard, one may use the modular squaring function in role of the trapdoor permutation above (see [68, 7, 346, 152]). This yields a secure public-key encryption scheme (depicted in Figure 1.2) with efficiency comparable to that of RSA. Recall that RSA itself is not secure (as it employs a deterministic encryption algorithm), whereas Randomized RSA

---

[11] Hard-core predicates are defined in Section 1.3. Recall that by [182], every trapdoor permutation can be modified into one having a hard-core predicate.

private-key: Two primes p, q, each congruent to 3 (mod 4).

public-key: Their product $N \overset{\text{def}}{=} pq$.

encryption of message $x \in \{0, 1\}^n$:
1. Uniformly select $s_0 \in \{1, ..., N\}$.
2. For $i = 1, .., n+1$, compute $s_i \leftarrow s_{i-1}^2 \bmod N$ and $\sigma_i = \text{lsb}(s_i)$.
The ciphertext is (s_{n+1}, y), where $y = x \oplus \sigma_1 \sigma_2 \cdots \sigma_n$.

decryption of the ciphertext (r, y):

Let $d = 2^{-n} \bmod \phi(N)$ [precomputed, where $\phi(N) \overset{\text{def}}{=} (p-1)(q-1)$].
1. Let $s_1 \leftarrow r^d \bmod N$.
2. For $i = 1, ..., n$, compute $\sigma_i = \text{lsb}(s_i)$ and $s_{i+1} \leftarrow s_i^2 \bmod N$.
The plaintext is $y \oplus \sigma_1 \sigma_2 \cdots \sigma_n$.

(One may think of n as being equal the length of N, but this is not essential to the scheme. The larger n, the more efficient the scheme becomes. Recall, however, that the security of the scheme depends on the length of N (and not on n).)

Fig. 1.2. The Blum–Goldwasser Public–Key Encryption Scheme [68].

(defined above) is not known to be secure under standard assumption such as intractability of factoring (or of inverting the RSA function).[12]

1.5.3 Security Beyond Passive Attacks

The above definitions refer only to a "passive" attack in which the adversary merely eavesdrops on the line over which ciphertexts are being sent. Stronger types of attacks, in which the adversary is active, may be possible in various applications. In particular, we discuss two notions of Chosen Ciphertext Attacks. In the first type (cf., [280]), the adversary may obtain the plaintexts of ciphertexts of its choice (as well as ciphertexts of plaintexts of its choice), and its task is to obtain information about the plaintext of a different ciphertext (to be presented in the future). In the second and stronger notion (cf., [307]), the adversary is given a target ciphertext ahead of time, and may obtain the plaintexts of any *other* ciphertext of its choice.

The private-key encryption scheme based on pseudorandom functions (described above) is secure also against Chosen Ciphertext Attacks of the first type. Public-key encryption schemes secure against Chosen Ciphertext Attacks of the first type can be constructed, assuming the existence of trap-door permutations and utilizing non-interactive zero-knowledge proofs [280] (which can be constructed under this assumption [143]). Public-key encryption schemes secure against Chosen Ciphertext Attacks of the second type are exactly those which are non-malleable (see below) under such attacks.

---

[12] Recall that Randomized RSA is secure assuming that the $n/2$ least significant bits constitute a hard-core function for n-bit RSA moduli. We only know that the $O(\log n)$ least significant bits constitute a hard-core function for n-bit moduli [7].

Loosely speaking, an encryption scheme is *non-malleable* if it is infeasible for an adversary, given a ciphertext, to produce a valid ciphertext for a related plaintext [122]. That is, the adversary is deemed successful if it produces a certain ciphertext, regardless of whether it knows to which plaintext it corresponds or not. In case of public-key encryption, non-malleability implies security in the sense discussed above. Non-malleability also comes in several flavors corresponding to what the adversary may obtain prior to attempting to produce a violating ciphertext. We focus on the strongest notion of Chosen Ciphertext Attack, where the adversary may obtain the plaintext of any ciphertext of its choice (as long as it is different that the target ciphertext given ahead of time). It is easy to turn any secure private-key encryption scheme into one which is secure and non-malleable under such Chosen Ciphertext Attacks, by using a message authentication scheme on top of the encryption (i.e., authenticate the ciphertext using a MAC).[13] Analogous (secure and non-malleable under such attacks) public-key encryption schemes are known to exist assuming the existence of trapdoor permutation [122].

For a detailed discussion of the relationship among the various notions of secure encryption the reader is referred to [35].

1.6 Signatures

Again, there are private-key and public-key versions both consisting of three efficient algorithms: *key generation, signing* and *verification*. (Private-key signature schemes are commonly referred to as *message authentication schemes* or *codes* (MAC).) The difference between the two types is again reflected in the definition of security (i.e., whether the adversary is given access to the verification-key). This difference yields different functionality (even more than in the case of encryption): Public-key signature schemes (hereafter referred to as signature schemes) may be used to produce signatures which are *universally verifiable* (given access to the publicly available verification-key of the signer). Private-key signature schemes (hereafter referred to as message authentication schemes) are only used to authenticate messages sent among a small set of *mutually trusting* parties (since ability to verify signatures is linked to the ability to produce them). Put in other words, message authentication schemes are used to authenticate information sent between (typically two) parties, and the purpose is to convince *the receiver* that the information was indeed sent by the legitimate sender. In particular, message authentication schemes cannot convince *a third party* that the sender has indeed sent the information (rather than the receiver having generated it by itself). In contrast, public-key signatures can be used to convince third parties: A signature to a document is typically sent to a second party so that in the future this party may (by merely presenting the signed document) convince

---

[13] See the definition of message authentication in the next section.

third parties that the document was indeed generated/sent/approved by the signer.

1.6.1 Definitions

We consider very powerful attacks on the signature scheme as well as a very liberal notion of breaking it. Specifically, the attacker is allowed to obtain signatures to any message of its choice. One may argue that in many applications such a general attack is not possible (as messages to be signed must have a specific format). Yet, our view is that it is impossible to define a general (i.e., application-independent) notion of admissible messages, and thus a general/robust definition of an attack seems to have to be formulated as suggested here. (Note that at worst, our approach is overly cautious.) Likewise, the adversary is said to be successful if it can produce a valid signature to ANY message for which it has not asked for a signature during its attack. Again, this defines the ability to form signatures to possibly "nonsensical" messages as a breaking of the scheme. Yet, again, we see no way to have a general (i.e., application-independent) notion of "meaningful" messages (so that only forging signatures to them will be consider a breaking of the scheme).

Definition 1.9 (unforgeable signatures [200]):

- *A* chosen message attack *is a process which on input a verification-key can obtain signatures* (relative to the corresponding signing-key) *to messages of its choice.*
- *Such an attack is said to* succeeds (in existential forgery) *if it outputs a valid signature to a message for which it has* NOT *requested a signature during the attack.*
- *A signature scheme is* secure (or unforgeable) *if every* feasible *chosen message attack succeeds with at most negligible probability.*

We stress that *plain* RSA (alike plain versions of Rabin's scheme [303] and DSS [271]) is not secure under the above definition. However, it may be secure if the message is "randomized" before RSA (or the other schemes) is applied (cf., [52]). Thus, the randomization paradigm (see Section 1.5) seems pivotal here too.

1.6.2 Constructions

Message authentication schemes can be constructed using pseudorandom functions (see [175] or the more efficient constructions in [47, 44, 32]). However, as noted in [33], an *extensive* usage of pseudorandom functions seem an overkill for achieving message authentication, and more efficient schemes may be obtained based on other cryptographic primitives. We mention two approaches, each consisting of a two-stage process:

1. *Fingerprinting* the message using a scheme which is *secure against forgery provided that the adversary does not have access to the scheme's outcome* (e.g., using Universal Hashing [92]), and *"hiding"* the result using a *non-malleable* scheme (e.g., a private-key encryption or a pseudorandom function). (Non-malleability is not required in certain cases; cf. [348, 236].)

2. *Hashing* the message *using a collision-free scheme* (cf., [115, 116]), and *authenticating* the result using a MAC which operates on (short) fixed-length strings [33].

Three central paradigms in the construction of *signature schemes* are the "refreshing" of the "effective" signing-key, the usage of an "authentication tree" and the "hashing paradigm".

The refreshing paradigm [200]: To demonstrate this paradigm, suppose we have a signature scheme which is robust against a "random message attack" (i.e., an attack in which the adversary only obtains signatures to uniformly distributed messages). Further suppose that we have a *one-time* signature scheme (i.e., a signature scheme which is secure against an attack in which the adversary obtains a signature to a single message of its choice). Then, we can obtain a secure (full-fledged) signature scheme as follows: When a new message is to be signed, we generate a new random signing-key for the one-time signature scheme, use it to sign the message, and sign the corresponding (one-time) verification-key using the fixed signing-key of the main signature scheme[14] (which is robust against a "random message attack") [131]. We note that one-time signature schemes (as utilized here) are easy to construct (see, for example [262]).

The tree paradigm [261, 200]: To demonstrate this paradigm, we show how to construct a general signature scheme using only a one-time signature scheme (alas one where an $2n$-bit string can be signed w.r.t an n-bit long verification-key). The idea is to use the initial singing-key (i.e., the one corresponding to the public verification-key) in order to sign/authenticate two new/random verification keys. The corresponding signing keys are used to sign/authenticate four new/random verification keys (two per a signing key), and so on. Stopping after d such steps, this process forms a binary tree with 2^d leaves where each leaf corresponds to an instance of the one-time signature scheme. The signing-keys at the leaves can be used to sign the actual messages, and the corresponding verification-keys may be authenticated using the path from the root. Pseudorandom functions may be used to eliminate the need to store the values of intermediate vertices used in previous signatures [165]. Employing this paradigm and assuming that the RSA function is infeasible to invert, one obtains a secure signature scheme [200, 165] in which

---

[14] One may generate the one-time key-pair and the signature to its verification-key ahead of time, leading to an "off-line/on-line" signature scheme [131]. An alternative and more efficient transformation, of signature schemes which are robust under a "random message attack" into general ones, has been suggested in [113].

the ith message can be signed/verified in time $2\log_2 i$ slower than plain RSA. Using a tree of large fan-in and assuming that RSA is infeasible to invert, one may obtain a secure signature scheme [127, 110] which for reasonable parameters is only 5 times slower than plain RSA.[15] We stress that plain RSA is not a secure signature scheme, whereas the security of its randomized version (mentioned above) is not known to be reducible to the assumption that RSA is hard to invert.

The hashing paradigm: It is common practice to sign real documents via a two stage process: First the document is hashed into a (relatively) short bit string, and next the basic signature scheme is applied to the resulting string. We note that this heuristic becomes sound provided the hashing function is *collision-free* (as defined in [115]). Collision-free functions can be constructed assuming the intractability of factoring [115]. One may indeed postulate that certain off-the-shelf products (as MD5 or SHA) are collision-free, but such assumptions need to be tested (and indeed may turn out false). We stress that using a hashing scheme in the above two-stage process without evaluating whether it is collision-free is a very dangerous practice.

A useful variant on the above paradigm is the use of *Universal One-Way Hash Functions* (as defined in [279]), rather than the collision-free hashing used above. In such a case a new hash function is selected per each application of the scheme, and the basic signature scheme is applied to both the (succinct) description of the hash function and to the resulting (hashed) string. (In contrast, when using a collision-free hashing function, the same function – the description of which is part of the signer's public-key – is used in all applications.) The advantage of using Universal One-Way Hash Functions is that their security requirement seems weaker than the collision-free condition (e.g., the former may be constructed using any one-way function [313], whereas this is not known for the latter).

A plausibility result [279, 313]: Signature schemes exist if (and only if) one-way functions exist. Unlike the constructions of signature schemes described above, the known construction of signature schemes from *arbitrary* one-way functions has no practical significance [313]. It is indeed an important open problem to provide an alternative construction which may be practical and still utilize an *arbitrary* one-way function.

1.6.3 Two Variants

Loosely speaking, Fail-stop signatures (cf., [296]) are signature schemes augmented by a proof system which allows the signer to prove that a particular (document,signature)-pair was not generated by him/her. In particular, each

---

[15] This figure refers to signing up-to 1,000,000,000 messages. The scheme in [127] requires a universal set of system parameters consisting of 1000–2000 integers of the size of the moduli. In the [110] scheme this requirement is removed.

document has many possible valid signatures (with respect to the public verification key), but only a negligible fraction of these can be generated by the unknown signing key. Furthermore, any strategy (even a non-computable one), is unlikely to generate signatures corresponding to the signing-key, and it is infeasible given one signing-key to generate valid signatures (i.e., w.r.t the verification key) which do not correspond to the given signing-key. Thus, fail-stop signature schemes allow to prove that forgery has occurred, and so offer an information-theoretic security guarantee to the potential signers (yet the guarantee to potential signature-recipients is only a computational one).[16] Furthermore, in case a proof of forgery is ever presented, one may "discertify" the particular verification key, or even the entire signature scheme (hence the term "fail-stop").

Loosely speaking, Blind signatures (cf., [96, 156, 299, 224]) are a variant of signature schemes in which the signer *gains no knowledge about the document it has signed*, but rather only knows the total number of documents signed. The unforgeability condition thus requires that it is infeasible to produce more signatures than the count held by the signer (and that this count reflects the number of successfully-completed invocations of the signing protocol). Blind signatures play a central role in the design of electronic cash systems (cf., [96, 98]): They are used to make the monetary-certificates, signed by a financial institute, untraceable.

1.7 Cryptographic Protocols

A general framework for casting cryptographic (protocol) problems consists of specifying a random process which maps n inputs to n outputs. The inputs to the process are to be thought of as local inputs of n parties, and the n outputs are their corresponding local outputs. The random process describes the desired functionality. That is, if the n parties were to trust each other (or trust some outside party), then they could each send their local input to the trusted party, who would compute the outcome of the process and send each party the corresponding output. The question addressed in this section is to what extent can this trusted party be "emulated" by the mutually distrustful parties themselves.

---

[16] The above refers to the natural convention by which a proof of forgery frees the signer of any obligations implied by the document. Thus, when accepting a valid signature the recipient is only guaranteed that it is infeasible for the signer to repudiate the signature. In contrast, when following the standard paradigms of signature schemes, the signature recipients have an absolute guarantee; whenever the verification algorithm accepts a signature, it is by definition an unrepudiated one.

1.7.1 Definitions

For simplicity we consider the special case where the specified process is deterministic and the n outputs are identical. That is, we consider an arbitrary n-ary function and n parties which wish to obtain the value of the function on their n corresponding inputs. Each party wishes to obtain the correct value of the function and prevent any other party from gaining anything else (i.e., anything beyond the value of the function and what is implied by it). Towards this goal, the parties execute a "secure" multi-party protocol.

We first observe that each party may change its local input before entering the protocol. However, this is unavoidable even when the parties utilize a trusted party. In general, the basic paradigm underlying the definitions of *secure multi-party computations* amounts to saying that situations which may occur in the real protocol, can be simulated in the ideal model (where the parties may employ a trusted party). Thus, the "effective malfunctioning" of parties in secure protocols is restricted to what is postulated in the corresponding ideal model. The specific definitions differ in the specific restrictions and/or requirements placed on the parties in the real computation. This is typically reflected in the definition of the corresponding ideal model – see examples below.

An example – computations with honest majority: Here we consider an ideal model in which any minority group (of the parties) may collude as follows. Firstly this minority shares its original inputs and decided together on replaced inputs[17] to be sent to the trusted party. (The other parties send their respective original inputs to the trusted party.) When the trusted party returns the output, each majority player outputs it locally, whereas the colluding minority may compute outputs based on all they know (i.e., the output and all the local inputs of these parties). A *secure multi-party computation with honest majority* is required to emulate this ideal model. That is, the effect of any feasible adversary which controls a minority of the players in the actual protocol, can be essentially simulated by a (different) feasible adversary which controls the corresponding players in the ideal model. This means that in a secure protocol the effect of each minority group is "essentially restricted" to replacing its own local inputs (independently of the local inputs of the majority players) before the protocol starts, and replacing its own local outputs (depending only on its local inputs and outputs) after the protocol terminates. (We stress that in the real execution the minority players do obtain additional pieces of information; yet in a secure protocol they gain nothing from these additional pieces of information, as they can actually reproduce these by themselves.)

---

[17] Such replacement may be avoided if the local inputs of parties are verifiable by the other parties. In such a case, a party (in the ideal model) has the choice of either joining the execution of the protocol with its correct local input or not join the execution at all (but it cannot join with a replaced local input). Secure protocols emulating this ideal model can be constructed as well.

Secure protocols according to the above definition may even tolerate a situation where a minority of the parties aborts the execution. An aborted party (in the real protocol) is simulated by a party (in the ideal model) which aborts the execution either before supplying its input to the trusted party (in which case a default input is used) or after supplying its input. In either case, the majority players (in the real protocol) are able to compute the output although a minority aborted the execution. This cannot be expected to happen when there is no honest majority (e.g., in a two-party computation) [107].

Another example – two-party computations: In light of the above, we consider an ideal model where each of the two parties may "shut-down" the trusted (third) party at any point in time. In particular, this may happen after the trusted party has supplied the outcome of the computation to one party but before it has supplied it to the second. A *secure multi-party computation allowing abort* is required to emulate this ideal model. That is, each party's "effective malfunctioning" in a secure protocol is restricted to supplying an initial input of its choice and aborting the computation at any point in time. We stress that, as above, the choice of the initial input of each party may NOT depend on the input of the other party.

1.7.2 Constructions

General plausibility results: Assuming the existence of trapdoor permutations, one may provide secure protocols for ANY two-party computation (allowing abort) [353] as well as for ANY multi-party computations with honest majority [186]. Thus, a host of cryptographic problems are solvable assuming the existence of trapdoor permutations. Specifically, any desired (input–output) functionality can be enforced, provided we are either willing to tolerate "early abort" (as defined above) or can rely on a majority of the parties to follow the protocol. Analogous plausibility results were subsequently obtained in a variety of models. In particular, we mention secure computations in the private channels model [59, 97], in the presence of mobile adversaries [293], and for an adaptively chosen set of corrupted parties [86].

As stressed in the case of zero-knowledge proofs, we view these results as asserting that very wide classes of problems are solvable in principle. However, we do not recommend using the solutions derived by these general results in practice. For example, although Threshold Cryptography (cf., [120, 160]) is merely a special case of multi-party computation, it is indeed beneficial to focus on its specifics.

1.8 Some Notes

We partition the notes into two categories: *General notes* which refer to general themes in this chapter, and *specific notes* which refer to specific covered or uncovered issues.

1.8.1 General Notes

On information theoretic secrecy. Most of Modern Cryptography aims at achieving *computational* security; that is, making it infeasible (rather than impossible) for an adversary to break the system. The departure from *information theoretic* secrecy was suggested by Shannon in the very paper which introduced the notion [323]: In an information theoretic secure encryption scheme the private-key must be longer than the total entropy of the plaintexts to be sent using this key. This drastically restricts the applicability of (information-theoretic secure) private-key encryption schemes. Furthermore, notions such as public-key cryptography, pseudorandom generators, and most known cryptographic protocols[18] cannot exist in an information theoretic sense.

On the need for and choice of assumptions. As stated in Section 1.2, most of Modern Cryptography is based on computational difficulty. Intuitively, this is an immediate consequence of the fact that Modern Cryptography wish to capitalize on the difference between feasible attacks and possible-but-infeasible attacks. Formally, the existence of one-way functions has been shown to be a necessary condition for the existence of secure private-key encryption [218], pseudorandom generators [245], digital signatures [313], "non-trivial" zero-knowledge proofs [292], and various basic protocols [218].

As we need assumptions anyhow, why not assume what we want? Well, first we need to know what we want. This calls for a clear definition of complex security concerns – an non-trivial issue which is discussed at length in previous sections. However, once a definition is derived how can we know that it can at all be met? The way to demonstrate that a definition is viable (and so the intuitive security concern can be satisfied at all) is to construct a solution based on a *better understood* assumption. For example, looking at the definition of zero-knowledge proofs [199], it is not a-priori clear that such proofs exists in a non-trivial sense. The non-triviality of the notion was demonstrated in [199] by presenting a zero-knowledge proof system for statements, regarding Quadratic Residuosity, which are believed to be hard to verify (without extra information). Furthermore, in contrary to prior beliefs, it was shown in [185] that the existence of commitment schemes[19] implies that any NP-statement can be proven in zero-knowledge. Thus, facts which were not known at all to hold (and even believed to be false), where shown to hold by reduction to widely believed assumptions (without which most

---

[18] Here we refer to cryptographic protocols in the "standard model" where the adversary can read all messages sent between honest parties. In contrast, information-theoretically secure multi-party computation is possible when assuming the existence of perfect private channels between each pair of honest users [59, 97].

[19] Consequently, it was shown how to construct commitment schemes based on any pseudorandom generator [272], and that the latter exists if one-way functions exist [211].

of Modern Cryptography collapses anyhow). Furthermore, reducing the solution of a new task to the assumed security of a well-known primitive typically means providing a construction which, using the known primitive, solves the new task. This means that we do not only know (or assume) that the new task is solvable but rather have a solution based on a primitive which, being well-known, typically has several candidate implementations. More on this subject below.

On the meaning of asymptotic results. Asymptotic analysis is a major simplifying convention. It allows to disregard specifics like the model of computation and to focus on the essentials of the problem at hand. Further simplification is achieved by identifying efficient computations with polynomial-time computations, and more importantly by identifying infeasible computations with ones which are not implementable in polynomial-time. However, none of these conventions is really essential for the theory discussed in this chapter.[20]

As stated in Section 1.2, all know results (referring to computational complexity) consists of an explicit construction in which a complex primitive is implemented based on a simpler one. The claim of security in many papers merely states that if the resulting (complex) primitive can be broken in polynomial-time then so can the original (simpler) primitive. However, all papers provide an explicit construction showing how to use any breaking algorithm for the resulting primitive in order to obtain a breaking algorithm for the original primitive. This transformation does not depend on the running-time of the first algorithm; it typically uses the first algorithm as a black-box. Thus, the running-time of the resulting breaking algorithm (for the simpler primitive) is explicitly bounded in terms of the running-time of the given breaking algorithm (for the complex primitive). This means that for each of these results, one can instantiate the resulting (complex) scheme for any desired value of the security parameter, make a concrete assumption regarding the security of the underlying (simpler) primitive, and derive a concrete estimate of the security of the proposed implementation of the complex primitive.

The applicability of a specific theoretical result depends on the complexity of the construction and the relation between the security of the resulting scheme and the quantified intractability assumption. Some of these results seem applicable in practice, some only offer useful paradigm/techniques, and other only state the plausibility of certain results. In the latter cases it is indeed the task of the theory community to work towards the improvement of these results. In fact, many improvements of this type have been achieved in the past (and we hope to see more in the future). Following are some examples:

---

[20] As long as the notions of efficient and feasible computation are sufficiently robust and rich. For example, they should be closed under various functional compositions and should allow computations such as RSA.

- A plausibility result of Yao (commonly attributed to [351]) on the existence of hard-core predicates, assuming the existence of one-way permutations, was replaced by a practical construction of hard-core predicates for any one-way functions [182].
- A plausibility result of Yao (commonly attributed to [351]) by which any weak one-way permutation can be transformed into an ordinary one-way permutation was replaced by an efficient transformation of weak one-way permutation into ordinary one-way permutation [177].
- A plausibility result of [185] by which one may construct Verifiable Secret Sharing schemes (cf., [106]), using any one-way function, was replaced by an efficient construction the security of which is based on DLP [147]. In general, many concrete problems which are solvable in principle (by the plausibility results of [185, 353, 186]) were given efficient solutions.

Forget the result, use its ideas. As stated above, some theoretical results are not directly applicable in practice. Still, in many cases these results utilize ideas which may be of value in practice. Thus, if you know (by a theoretical result) that a problem is solvable in principle, but the known construction is not applicable for your purposes, you may try to utilize some of its underlying ideas when trying to come-up with an alternative solution tailored for your own purposes. We note that the underling ideas are at least as likely to appear in the proof of security as in the construction itself.

The choice of assumptions, revisited. When constructing a solution to a cryptographic problem one may have a choice of which building blocks to use (e.g., one-way functions or pseudorandom functions). In a coarse sense these tools may look equivalent (e.g., one exists if and only if the other exists), but when deciding which to use in practice one should consider the actual level of security attributed to each of them and the "cost" of using each of them as a building block in a particular construction. In the latter term ("cost") we mean the relationship of the security of the building block to the security of the resulting solution. For further discussion the reader is referred to [32, Sec. 1.5]. *Turning the table around*, if we note that a specific primitive provides good security, when used as a building block in many constructions, then this may serve as incentive to focus attention on the implementation of this primitive. The last statement should be understood both as referring to the theory and practice of cryptography. For example, it is our opinion that the industry should focus on constructing fixed-length-key pseudorandom functions rather than on constructing fixed-length-key pseudorandom permutations (or, equivalently, private-key block ciphers).[21]

Security as a quantity rather than a quality. From the above it should be clear that our notions of security are quantitative in nature. They refers to the minimal amount of work required to break the system (as a function

---

[21] Not to mention that the latter can be efficiently constructed from the former [252, 277].

of the security parameter). Thus alternative constructions for the same task may (and need to) be compared based on the security they provide. This can be done whenever the underlying assumption are comparable.

"Too cautious" definitions. As stated in Sections 1.5 and 1.6, our definitions of security seem "too cautious" in the sense that they also imply things which may not matter in practice. This is an artifact of our approach to security which requires that the adversary gains *nothing* (rather than "gains nothing we care about") by its malicious actions. We stress two advantages of our approach. First it yields application-independent notions of security (since the notion of a "gain we care about" is application-dependent). Secondly, even when having a specific application in mind, it is close to impossible to come-up with a precise characterization of the set of "gains we care about". Thus, even in the latter case, our approach of depriving the adversary from *any* gain seems to be the best way to go. Finally, we note that in all known cases the plausibility of meeting the "maximalistic" definitions of security has been demonstrated (based on assumptions which are necessary even for "minimalistic" notions of security).

On "Provable Security". Some of the papers discussed in this chapter use the term "provable security". The term is supposed to reflect the fact that these papers only make well-defined technical claims and that proofs of these claims are given or known to the authors. Specifically, whenever a term such as "security" is used, the paper offers or refers to a rigorous definition of the term (and the authors wish to stress this fact in contrast to prior papers where the term was used as an undefined intuitive phrase). We personally object to this terminology since it suggests the possibility that there can be technical claims[22] which are well-defined and others which are not, and among the former some can be stated even when no proof is known. This view is wrong: A technical claim must always be well-defined, and it must always have a proof (otherwise it is a conjecture – not a claim). There is room for non-technical claims, but these claims should be stated as opinions and such. In particular, a technical claim referring to security must always refer to a rigorous definition of security and the person making this claim must always know a proof (or state the claim as a conjecture).

Still, do consider specific attacks (but as a last resort). We do realize that sometimes one is faced with a situation where all the paradigms described above offer no help. A typical example occurs when designing an "atomic" cryptographic primitive (e.g., a one-way function). The first thing we suggest in such a case is to formulate precise specifications/assumptions regarding the security of this primitive. Once this is done, one may need to turn to ad-hoc methods for trying to test these assumptions (i.e., if the known attack schemes fail then one gains some confidence in the validity of the assumptions). For example, if we were to invent RSA today then we would

---

[22] We refer to theorems, lemmas, propositions and such.

have postulated that it is a trapdoor permutation. To evaluate the validity of our conjecture, we would have noted (as Rivest, Shamir and Adleman did in [312, Sec. IX]) that known algorithms for factoring are infeasible for reasonable values of the security parameter, and that there seems to be no other way to invert the function.

1.8.2 Specific Notes

This chapter can not possibly cover all good work done in Cryptography, not even all good work of theoretical flavor, and not even all theoretical work which interests the author. We have focused on one fundamental research direction – the attempt to turn Cryptography from an art into a science. Furthermore, within this direction we have preferred to concentrate on the basics, and gave-up on many important developments which go beyond the basics. Whenever such developments are mentioned it is typically in order to demonstrate a basic paradigm. Thus, the choice of material is governed by its relevance to the intentions of the current chapter. Arguably and hopefully this is correlated with the importance of the work, but no tight relation was sought or is claimed. In an attempt to redeem some of the omissions made above, we shortly discuss some topics which were ignored (or mentioned too briefly) above. The following collection of notes is indeed eclectic in nature.

Information theoretic secrecy, revisited. As stated above, most of Modern Cryptography only aims at achieving computational secrecy – and does so for a good reason (as information theoretic secrecy is unachievable in many settings). However, these impossibility results hold only in case the adversary has full information (apart from the honest parties secret inputs and private coin tosses). For example, information-theoretically secure multi-party computation is possible (and in fact feasible) if there are perfect private channels between each pair of honest users [59, 97]. On the other hand, information-theoretically secure private channels can be implemented on top of channels to which the adversary has limited access. Channels of the latter type include

1. *Quantum Channels* where an adversary is prevented from obtaining full information by the laws of quantum mechanics (cf., [76] and the references therein).
2. The *noisy channel model* (which generalizes the *wiretap channel* of [350]) where both the communication between the legitimate parties and the tapping channel of the adversary are subjected to noise (cf., [259, 114] and the references therein).
3. A model where the adversary can freely tap the communication channel but is restricted in the amount of data it can store (cf., [80]).

In addition, with respect to private-key cryptography (i.e., both encryption and message-authentication), the abovementioned impossibility results may be irrelevant in some applications. What these impossibility results actually establish is that the private-keys need to be at least as long as the data to which they are applied. In certain cases, especially given current storage technology, using such long private-keys may be feasible.

Byzantine Agreement. The general results regarding multi-party computations surveyed in Section 1.7 assume the existence of a *broadcast channel* (i.e., a channel on which each party may place messages which may be read by all parties and yet cannot be corrupted by any party). Such a channel can be implemented over a standard point-to-point network using a Byzantine Agreement protocol [295]. Efficient Byzantine Agreement protocols are known in a variety of models. In the information-theoretic model, we mention the deterministic protocols of [123, 338] which tolerate malicious behavior of $t < m/3$ parties, where m is the total number of parties. In the computational model, using a signature infrastructure, one may construct efficient protocols tolerate any number of faults [124]. These protocols operate in $O(t)$ rounds, which is optimal (for deterministic protocols). Assuming the existence of private channels, a faster (i.e., expected constant number of rounds) *randomized* algorithm tolerating $\Omega(m)$ malicious parties is known [148].

Threshold Cryptography. Cryptography relies on the user's ability to maintain the secrecy of its private-keys. However, guaranteeing the secrecy of private-keys in practice is not easy, especially when these keys belong to large organizations. It is thus desirable to replace the single private-key by a set of "shares" so that the disclosure of a *small* subset of shares *does not endanger* the *security* of the system, whereas a *larger* subset of shares *enables* the *operation* of the system. Assuming these shares are stored at different sites (and that after set-up time the private-key is never available again in any single site), such a scheme may enhance security as it seems harder to penetrate to several sites than to one. The security and operation of such a distributed cryptographic system falls within the domain of general multi-party computation, and thus is solvable in principle [186, 59, 97].[23] However, what one desires is efficient solutions, and in particular ones comparable in efficiency to standard "single private-key" cryptosystems. Such efficient solutions, called *threshold cryptosystems*, were envisioned in [119, 120] and provided in [120, 118, 162] (and many other works). In addition to the conditions informally described above, it is desired that the threshold system be *robust* [162] and *proactive* [293, 91, 214]. By robust we mean that proper operation is guaranteed even if some of the sites holding shares of the private-key misbehave (as may be the case when controlled by an adversary). By proactive we mean that both security and proper operation are maintained even if the adversary can, during the lifetime of the system, *gain temporary control of each site* provided it *never controls simultaneously a large number of sites*.

---

[23] Such a solution would use a standard secret-sharing scheme, and consists of "emulating" the reconstruction and usage of the private-key by an ideal trusted party, without having the key actually reconstructed in any site.

On the Random Oracle Model. A popular methodology for designing cryptographic protocols consists of the following two steps. One first designs an *ideal* system in which all parties (including the adversary) have oracle access to a truly random function, and proves the security of this ideal system. Next, one replaces the random oracle by a "good cryptographic hashing function" (such as MD5 or SHA), providing all parties (including the adversary) with the succinct description of this function. Thus, one obtains an *implementation* of the ideal system in a world where random oracles do not exist. This methodology, explicitly formulated in [49], has been used in many works (see, for example, [150, 320, 52]). However, it is unclear to what extent this methodology can be put on firm grounds. In particular, there exist secure ideal encryption and signature schemes, which do not have *any* secure implementation (cf., [88]). Thus, one cannot hope to "implement" (by a function ensemble) all properties of a random oracle. Instead, we suggest that one should proceed by identifying useful *special-purpose* properties of a random oracle, which can be also provided by a fully specified function (or function ensemble), and so yield implementations of certain useful ideal systems. In fact, first steps in this alternative direction have been taken in [82, 89].

1.9 Historical Perspective

Work done during the 1980's plays a dominant role in our exposition. This work was in turn tremendously influenced by previous work, but these influences were not stated explicitly above. The influence took the form of setting intuitive goals, providing basic techniques, and suggesting potential solutions which served as a basis for constructive criticism (leading to robust approaches). In this section, specifically in its first part, we try to trace some of these influences. We then proceed to the history of the rigorous (or robust) approach to cryptography.

Classic Cryptography. Answering the fundamental question of classic cryptography in a gloomy way (i.e., it is *impossible* to design a code that cannot be broken), Shannon also suggested a modification to the question [323]: Rather than asking whether it is *possible* to break the code, one should ask whether it is *feasible* to break it. A code should be considered good if it cannot be broken when investing work which is in reasonable proportion to the work required of the legal parties using the code. Indeed, this is the approach followed by Modern Cryptography.

New Directions in Cryptography. Prospects of commercial application were the trigger for the beginning of civil investigations of encryption schemes. The DES designed in the early 70's has adopted the new paradigm: It is clearly *possible*, but supposedly *infeasible* to break it. Following the challenge of constructing and analyzing new (private-key) encryption schemes, came new questions like how to exchange keys over an insecure channel [260].

New concepts were invented: *digital signatures* (cf., Diffie and Hellman [121] and Rabin [302]), *public-key cryptosystems* [121], and *one-way functions* [121]. First implementations of these concepts were suggested by Merkle and Hellman [264], Rivest, Shamir and Adleman [312], and Rabin [303].

Cryptography was explicitly related to complexity theory in [75, 133, 242]: It was understood that problems related to breaking a 1-1 cryptographic mapping cannot be $\mathcal{NP}$-complete, and more importantly that $\mathcal{NP}$-hardness of the breaking task is a poor evidence for cryptographic security. Techniques such as "*n*-out-of-2*n* verification" [302] and secret sharing [324] were introduced (and indeed were used extensively in subsequent research).

At the Dawn of a New Era. Early investigations of cryptographic protocols revealed the inadequacy of imprecise notions of security and the subtleties involved in designing cryptographic protocols. In particular, problems as *coin tossing over telephone* [62], *exchange of secrets* [61], and *Oblivious Transfer* were formulated [304] (cf., [130]). Doubts (raised by Lipton) concerning the security of the "mental poker" protocol of [326] led to the current notion of secure encryption, due to Goldwasser and Micali [198], and to concepts as computational indistinguishability [198, 351]. Doubts (raised by Fischer) concerning the Oblivious Transfer protocol of [304] led to the concept of zero-knowledge (suggested by Goldwasser, Micali, and Rackoff [199], with early versions dating to March 1982).

A formal approach to the security of cryptographic protocols was suggested in [125]. This approach actually identifies a subclass of insecure protocols (i.e., those which can be broken via a syntactically-restricted type of attack). Furthermore, it turned out that it is much too difficult to test whether a protocol is secure [129]. Recall that, in contrast, our current approach is to construct secure protocols (alongside with their proof of security), and that this approach is *complete* (in the sense that it allows to solve any solvable problem).

Establishing the new paradigms. The abovementioned work of Goldwasser and Micali [198] is the key-stone of the rigorous approach to cryptography. On top of supplying robust definitions for secure encryption – the most classic of cryptographic tasks – it has introduced almost all paradigms which played a key role in subsequent developments. We refer firstly to the *simulation paradigm*, made more explicit in the definition of zero-knowledge [199], and to the notion of *computational indistinguishability*, formulated in full generality by Yao [351]. But not less importantly, we refer to the understanding that cryptographic tasks are highly complex entities which should be "reduced" to simpler ones (such as well-defined intractability assumptions referring to simply stated computational problems).

The next major step was the definition and construction of pseudorandom generators by Blum, Micali, and Yao [71, 351]. In addition to the contribution of these works to cryptography, they have established a link between cryptography and computer science at large. This link, in turn, has fostered the evolution of cryptography from an art to a scientific discipline.

The concept of zero-knowledge, suggested by Goldwasser, Micali, and Rackoff [199], has provided an extremely powerful tool for the design of cryptographic protocols. In addition, it has further clarified the simulation paradigm, demonstrating its generality. Being such a fascinating notion, zero-knowledge has attracted attention also from outside of cryptography.

The above robust definitional approach would never have striven were it not coupled with actual constructions or at least proofs of feasibility. These were indeed provided – in many cases in the same papers (e.g., [198, 71, 351, 199]) and/or by subsequent work. The work of Goldreich, Micali, and Wigderson [185], which established the generality and wide applicability of zero-knowledge proofs, is a good example to the latter. We also mention the signature scheme of Goldwasser, Micali, and Rivest [200], which demonstrated – in contrary to prior beliefs – that a robust definition of unforgeable signature schemes can be materialized.

Going beyond the wildest dreams. Meeting the above robust definitions of security qualifies as going beyond the wildest dreams of most researchers of the time. It is safe to say that the works of Yao [353] and of Goldreich, Micali, and Wigderson [186] went beyond the wildest dreams of anybody. These works demonstrate that any (properly defined) cryptographic protocol problem can be solved in a meaningful sense.

And still, going on. The last paragraph may be read as an invitation to "close shop". This is certainly not the intention. As stated throughout this exposition, there is still much to be done (see, for example, Section 1.10). Indeed, in the years which have elapsed, many important works have been done. We merely mention the project of basing each cryptographic task on the minimal possible intractability assumption (the constructions of pseudorandom generator by Håstad, Impagliazzo, Levin and Luby [211] and signature schemes by Naor, Yung and Rompel [279, 313] are indeed the crown jewels of this project), and replacing feasibility claims by practical constructions (the hard-core predicate of Goldreich and Levin [182] is a good example).

1.10 Two Suggestions for Future Research

A very important direction for future research consists of trying to "upgrade" the utility of some of the constructions mentioned above. In particular, we have highlighted four plausibility results: two referring to the construction of pseudorandom generators and signature schemes and two referring to the construction of zero-knowledge proofs and multi-party protocols. For the former two results, we see no fundamental reason why the corresponding constructions can not be replaced by reasonable ones (i.e., providing very efficient constructions of pseudorandom generators and signature schemes based on *arbitrary* one-way functions). Furthermore, we believe that working towards this goal may yield new and useful paradigms (which may be applicable in practice regardless of these results). As for the latter general plausibility results (i.e., the construction of zero-knowledge proofs and multi-party protocols), here there seem to be little hope for a result which may both maintain the generality of the results in [185, 353, 186] and yield practical solutions for each specific task. However, we believe that there is work to be done towards the development of additional paradigms and techniques which may be useful in the construction of schemes for specific tasks.

Another very important direction is to provide results and/or develop techniques for guaranteeing that individually-secure cryptographic protocols remain secure when many copies of them are run in parallel and, furthermore, obliviously of one another.[24] Although some negative results are known [179], they only rule out specific approaches (such as the naive false conjecture that ANY zero-knowledge proof maintains its security when executed twice in parallel).

1.11 Some Suggestions for Further Reading

The intention of these suggestions is NOT to provide a scholarly account of the due credits but rather to provide sources for further reading. Thus, our main criteria is the readability of the text (not its novelity). The recommendations are arranged by subjects.

One-Way Functions, Pseudorandom Generators and Zero-Knowledge: For these, our favorite source is our own text [170].

Encryption Schemes: A good motivating discussion appears in [198]. For a definitional treatment of eavesdropping security, the reader is referred to the revised version of [170]. Further details on the constructions of public-key encryption schemes (sketched above) can be found in [198, 167] and [68, 7], respectively. For discussion of Non-Malleable Cryptography, which actually transcends the domain of encryption, see [122].

Signature Schemes: For a definitional treatment of *signature schemes* the reader is referred to [200] and [298]. Easy to understand constructions appear in [48, 131, 127, 110]. Variants on the basic model are discussed in [298] and in [96, 156, 299, 224]. For discussion of *message authentication schemes* (MACs) the reader in referred to [33].

General Cryptographic Protocols: This area is both most complex and most lacking of good expositions. For the least of all evil, we refer the reader to [173] which provides an exposition of the basic definitions and results, as well as detailed proofs for the latter. More advanced treatment can be found in [81, 83].

---

[24] This goal coincides with a general formulation of non-malleable cryptography, as introduced in [122].

New Directions: These include Realizing the Random Oracle Model [82, 88, 89], Session-Key Problems [50, 51, 34], Incremental Cryptography [38, 39], Coercibility [87, 84], sharing of cryptographic objects [120, 118, 160], Private Information Retrieval [105, 102, 238], Cryptanalysis by induced faults [72], and many others.

Acknowledgments

I am most grateful to Hugo Krawczyk for carefully reading and commenting on an early draft of this chapter.

Thanks also to Mihir Bellare, Gilles Brassard, Christian Cachin, Ran Canetti, Ronald Cramer, Cynthia Dwork, Shafi Goldwasser, Moni Naor and Birgit Pfitzmann for comments and corrections regarding previous versions of this chapter.

2. Probabilistic Proof Systems

A proof is whatever convinces me.

Shimon Even, answering a student's question
in his Graph Algorithms class (1978)

Summary – Various types of *probabilistic* proof systems have played a central role in the development of computer science in the last decade. In this chapter, we concentrate on three such proof systems — *interactive proofs*, *zero-knowledge proofs*, and *probabilistic checkable proofs* — stressing the essential role of randomness in each of them.

2.1 Introduction

The glory attached to the creativity involved in finding proofs, makes us forget that it is the less glorified procedure of verification which gives proofs their value. Philosophically speaking, proofs are secondary to the verification procedure; whereas technically speaking, proof systems are defined in terms of their verification procedures.

The notion of a verification procedure assumes the notion of computation and furthermore the notion of efficient computation. This implicit assumption is made explicit in the definition of $\mathcal{NP}$, in which efficient computation is associated with (deterministic) polynomial-time algorithms.

Definition 2.1 (NP-proof systems): *Let $S \subseteq \{0,1\}^*$ and $\nu : \{0,1\}^* \times \{0,1\}^* \mapsto \{0,1\}$ be a function so that $x \in S$ if and only if there exists a $w \in \{0,1\}^*$ such that $\nu(x,w) = 1$. If ν is computable in time bounded by a polynomial in the length of its first argument then we say that S is an NP-set and that ν defines an NP-proof system.*

Traditionally, NP is defined as the class of NP-sets. Yet, each such NP-set can be viewed as a proof system. For example, consider the set of satisfiable

Boolean formulae. Clearly, a satisfying assignment π for a formula ϕ constitutes an NP-proof for the assertion "ϕ is satisfiable" (the verification procedure consists of substituting the variables of ϕ by the values assigned by π and computing the value of the resulting Boolean expression).

The formulation of NP-proofs restricts the "effective" length of proofs to be polynomial in length of the corresponding assertions (since the running-time of the verification procedure is restricted to be polynomial in the length of the assertion). However, longer proofs may be allowed by padding the assertion with sufficiently many blank symbols. So it seems that NP gives a satisfactory formulation of proof systems (with efficient verification procedures). This is indeed the case if one associates efficient procedures with *deterministic* polynomial-time algorithms. However, we can gain a lot if we are willing to take a somewhat non-traditional step and allow *probabilistic* verification procedures. In particular,

- Randomized and interactive verification procedures, giving rise to *interactive proof systems*, seem much more powerful (i.e., "expressive") than their deterministic counterparts.
- Such randomized procedures allow the introduction of *zero-knowledge proofs* which are of great theoretical and practical interest.
- NP-proofs can be efficiently transformed into a (redundant) form which offers a trade-off between the number of locations examined in the NP-proof and the confidence in its validity (see *probabilistically checkable proofs*).

In all the abovementioned types of probabilistic proof systems, explicit bounds are imposed on the computational complexity of the verification procedure, which in turn is personified by the notion of a verifier. Furthermore, in all these proof systems, the verifier is allowed to toss coins and rule by statistical evidence. Thus, all these proof systems carry a probability of error; yet, this probability is explicitly bounded and, furthermore, can be reduced by successive application of the proof system.

Notational Conventions. When presenting a proof system, we state all complexity bounds in terms of the length of the assertion to be proven (which is viewed as an input to the verifier). Namely, polynomial-time means time polynomial in the length of this assertion. Note that this convention is consistent with the definition of NP-proofs.

Denote by `poly` the set of all integer functions bounded by a polynomial and by `log` the set of all integer functions bounded by a logarithmic function (i.e., $f \in \texttt{log}$ iff $f(n) = O(\log n)$). All complexity measures mentioned in the subsequent exposition are assumed to be constructible in polynomial-time.

Organization. We start by discussing interactive proofs (in Section 2.2), and then turn to zero-knowledge proofs (Section 2.3) and probabilistically checkable proofs – PCP (Section 2.4). Other types of probabilistic proof systems are discussed in Section 2.5. These include multi-prover interactive proofs (MIP), two types of computationally-sound proofs (i.e., arguments and

CS-proofs), non-interactive probabilistic proofs, and proofs of knowledge. We conclude with a comparison among the various types of proof systems (Section 2.6.1), a brief historical account (Section 2.6.2) and some open problems (Section 2.6.3).

2.2 Interactive Proof Systems

In light of the growing acceptability of randomized and distributed computations, it is only natural to associate the notion of efficient computation with probabilistic and interactive polynomial-time computations. This leads naturally to the notion of an interactive proof system in which the verification procedure is interactive and randomized, rather than being non-interactive and deterministic. Thus, a "proof" in this context is not a fixed and static object, but rather a randomized (dynamic) process in which the verifier interacts with the prover. Intuitively, one may think of this interaction as consisting of "tricky" questions asked by the verifier, to which the prover has to reply "convincingly". The above discussion, as well as the following definition, makes explicit reference to a prover, whereas a prover is only implicit in the traditional definitions of proof systems (e.g., NP-proofs).

2.2.1 Definition

Loosely speaking, an interactive proof is a game between a computationally bounded verifier and a computationally unbounded prover whose goal is to convince the verifier of the validity of some assertion. Specifically, the verifier is probabilistic polynomial-time. It is required that if the assertion holds then the verifier always accepts (i.e., when interacting with an appropriate prover strategy). On the other hand, if the assertion is false then the verifier must reject with probability at least $\frac{1}{2}$, no matter what strategy is being employed by the prover. A sketch of the formal definition is given in Item (1) below. Item (2) introduces additional complexity measures which can be ignored in first reading.

Definition 2.2 (Interactive Proof systems – IP [199]):

1. *An* interactive proof system *for a set* S *is a two-party game, between a* verifier *executing a probabilistic polynomial-time strategy* (denoted V) *and a* prover *which executes a computationally unbounded strategy* (denoted P), *satisfying*
   - Completeness: *For every* $x \in S$ *the verifier* V *always accepts after interacting with the prover* P *on common input* x.
   - Soundness: *For every* $x \notin S$ *and every potential strategy* P^*, *the verifier* V *rejects with probability at least* $\frac{1}{2}$, *after interacting with* P^* *on common input* x.

2. *For an integer function m, the complexity class $\mathcal{IP}(m(\cdot))$ consists of sets having an interactive proof system in which, on common input x, at most $m(|x|)$ messages are exchanged[1] between the parties.*

   For a set of integer functions, M, we let $\mathcal{IP}(M) \overset{\text{def}}{=} \bigcup_{m \in M} \mathcal{IP}(m(\cdot))$.

   Finally, $\mathcal{IP} \overset{\text{def}}{=} \mathcal{IP}(\text{poly})$.

In Item (1), we have followed the standard definition which specifies strategies for both the verifier and the prover. An alternative presentation only specifies the verifier's strategy while rephrasing the completeness condition as follows:

> *There exists a prover strategy P so that, for every $x \in S$, the verifier V always accepts after interacting with P on common input x.*

Arthur-Merlin games (a.k.a *public-coin* proof systems), introduced in [23], are a special case of interactive proofs, where the verifier must send the outcome of any coin it tosses (and thus need not send any other information). Yet, as shown in [203], this restricted case has essentially the same power as the general case (introduced in [199]). Thus, in the context of interactive proof systems, asking random questions is as powerful as asking "tricky" ones. Also, in some sources interactive proofs are defined so that two-sided error probability is allowed; yet, this does not increase their power [158]. See further discussion below.

We stress that although we have relaxed the requirements from the verification procedure, by allowing it to interact, toss coins and risk some (bounded) error probability, we did not restrict the validity of its assertions by assumptions concerning the potential prover. (This should be contrasted with the latter notions of proof systems, such as computationally-sound ones and multi-prover ones, in which the validity of the verifier's assertions depends on assumptions concerning the external proving entity.)

2.2.2 The Role of Randomness

Randomness is essential to the formulation of interactive proofs; if randomness is not allowed (or if it is allowed but zero error is required in the soundness condition) then interactive proof systems collapse to NP-proof systems. The reason being that, in case the verifier is deterministic, the prover can predict the verifier's part of the interaction. Thus it suffices to let the (modified) prover send the full transcript of the (original) interaction, and let the (modified) verifier check that the transcript is indeed valid and accepting (i.e.,

---

[1] We count the total number of messages exchanged regardless of the direction of communication. For example, interactive proof systems in which the verifier sends a single message answered by a single message of the prover corresponds to $\mathcal{IP}(2)$. Clearly, $\mathcal{NP} \subseteq \mathcal{IP}(1)$, yet the inclusion may be strict since the verifier may toss coins after receiving the prover's single message.

that the verifier messages match the original (deterministic) verifier strategy and that the transcript would have caused the original verifier to accept).[2]

The moral is that there is no point to interact with predictable parties which are also computationally weaker. (This moral represents the prover's point of view. Certainly, from the verifier's point of view it is beneficial to interact with the prover, since the latter is computationally stronger.)

2.2.3 The Power of Interactive Proofs

A simple example demonstrating the power of interactive proofs follows. Specifically, we present an interactive proof for proving that two graphs are not isomorphic[3]. It is not known whether such a statement can be proven via an NP-proof system.

Construction 2.3 (Interactive proof for Graph Non-Isomorphism [185]):

- Common Input: *A pair of graphs, $G_1 = (V_1, E_1)$ and $G_2 = (V_2, E_2)$. Suppose, without loss of generality, that $V_1 = \{1, 2, ..., |V_1|\}$, and similarly for V_2.*
- Verifier's first step (V1): *The verifier selects at random one of the two input graphs, and sends to the prover a random isomorphic copy of this graph. Namely, the verifier selects uniformly $\sigma \in \{1, 2\}$, and a random permutation π from the set of permutations over the vertex set V_σ. The verifier constructs a graph with vertex set V_σ and edge set*

$$E \stackrel{\text{def}}{=} \{\{\pi(u), \pi(v)\} : \{u, v\} \in E_\sigma\}$$

and sends (V_σ, E) to the prover.
- Motivating Remark: *If the input graphs are non-isomorphic, as the prover claims, then the prover should be able to distinguish (not necessarily by an efficient algorithm) isomorphic copies of one graph from isomorphic copies of the other graph. However, if the input graphs are isomorphic then a random isomorphic copy of one graph is distributed identically to a random isomorphic copy of the other graph.*
- Prover's step: *Upon receiving a graph, $G' = (V', E')$, from the verifier, the prover finds a $\tau \in \{1, 2\}$ so that the graph G' is isomorphic to the input graph G_τ. (If both $\tau = 1, 2$ satisfy the condition then τ is selected arbitrarily. In case no $\tau \in \{1, 2\}$ satisfies the condition, τ is set to 0). The prover sends τ to the verifier.*

---

[2] Probabilistic verifiers of zero soundness error are dealt with by fixing their coins to an arbitrary outcome, say the all-zero sequence.

[3] Two graphs, $G_1 = (V_1, E_1)$ and $G_2 = (V_2, E_2)$, are called *isomorphic* if there exists a 1-1 and onto mapping, ϕ, from the vertex set V_1 to the vertex set V_2 so that $\{u, v\} \in E_1$ if and only if $\{\phi(v), \phi(u)\} \in E_2$. The ("edge preserving") mapping ϕ, if existing, is called an *isomorphism* between the graphs.

– Verifier's second step (V2): *If the message, τ, received from the prover equals σ (chosen in Step V1) then the verifier outputs 1 (i.e., accepts the common input). Otherwise the verifier outputs 0 (i.e., rejects the common input).*

The verifier's strategy presented above is easily implemented in probabilistic polynomial-time. We do not known of a probabilistic polynomial-time implementation of the prover's strategy, but this is not required. The motivating remark justifies the claim that Construction 2.3 constitutes an interactive proof system for the set of pairs of non-isomorphic graphs. Recall that the latter is a co$\mathcal{NP}$-set (not known to be in $\mathcal{NP}$).

Interactive proofs are powerful enough to prove *any* coNP assertion (e.g., that a graph is not 3-colorable) [255]. Furthermore, the class of sets having interactive proof systems coincides with the class of sets that can be decided using a polynomial amount of work-space [325].

Theorem 2.4 (The IP Theorem [255, 325]): $\mathcal{IP} = \mathcal{PSPACE}$.

Recall that it is widely believed that $\mathcal{NP} \subset \mathcal{PSPACE}$. Thus, under this conjecture, interactive proofs are more powerful than NP-proofs.

Theorem 2.4, was established using algebraic methods (see proof sketch below). In particular, the following approach – unprecedented in complexity theory – was employed: In order to demonstrate that a particular set is in a particular class, an arithmetic generalization of the Boolean problem is presented, and (elementary) algebraic methods are applied to show that the arithmetic problem is solvable within the class. Interestingly, this technique "does not relativize" and, furthermore, yields results (e.g., $\mathcal{IP} = \mathcal{PSPACE}$) that are false relative to most oracles, providing a dramatic refutation of the "Random Oracle Hypothesis"; see [99].

Sketch of the Proof of Theorem 2.4

We first show that co$\mathcal{NP} \subseteq \mathcal{IP}$, by presenting an interactive proof system for the co$\mathcal{NP}$-complete set of non-satisfiable CNF formulae. Next we modify this proof system to obtain the ultimate theorem. The first part is due to Lund, Fortnow, Karloff and Nisan [255], but our entire presentation follows the one of Shamir [325], to which the proof of the second part is due.[4]

Arithmetization of Boolean (CNF) formulae: Given a Boolean (CNF) formula, we replace the Boolean variables by integer variables, OR-clauses by sums, and the top level conjunction by a product. Then we sum over all 0-1 assignments to these variables. For example, the Boolean formula

$$(x_3 \vee \neg x_5 \vee x_{17}) \wedge (x_5 \vee x_9) \wedge (\neg x_3 \vee \neg x_4)$$

---

[4] Some people, consider the proof in [327] to be simpler than the one presented in [325] (and below). We are not among them.

is replaces by the arithmetic expression

$$(x_3 + (1 - x_5) + x_{17}) \cdot (x_5 + x_9) \cdot ((1 - x_3) + (1 - x_4))$$

and the Boolean formula is non-satisfiable if and only if the sum of the arithmetic expression over $x_1, x_2, ..., x_{17} \in \{0, 1\}$ equals 0. Observe that the arithmetic expression is a low degree polynomial. Also observe that, in any case, the value of the arithmetic expression is bounded above by v^m, where v is the maximum number of variables in a clause, and m is the number of clauses.

Moving to a Finite Field: Whenever we check equality between two integers in $[0, M]$, it suffices to check equality mod q, where $q > M$. The benefit is that the arithmetic is now in a finite field (mod q) and so certain things are "nicer" (e.g., uniformly selecting a value). Thus, proving that a CNF formula is not satisfiable reduces to proving equality of the following form

$$\sum_{x_1=0,1} \cdots \sum_{x_n=0,1} \phi(x_1, ..., x_n) \equiv 0 \pmod{q}$$

where ϕ is a low degree multi-variant polynomial.

The construction: We strip off summations in iterations. In each iteration the prover is supposed to supply the polynomial representing the expression in one (currently stripped) variable. (By the above observation, this is a low degree polynomial and so has a short description.) The verifier checks that the polynomial (say, p) is of low degree, and that it corresponds to the current value (say, v) being claimed (i.e., $p(0)+p(1) \equiv v$). Next, the verifier randomly instantiates the variable, yielding a new value to be claimed for the resulting expression (i.e., $v \leftarrow p(r)$, for uniformly chosen $r \in GF(q)$). The verifier sends the uniformly chosen instantiation to the prover. (At the end of the last iteration, the verifier has a fully specified expression and can easily check it against the claimed value.)

Thus, the ith iteration is aimed at proving a claim of the form

$$\sum_{x_i=0,1} \cdots \sum_{x_n=0,1} \phi(r_1, ..., r_{i-1}, x_i, x_{i+1}, ..., x_n) \equiv v_{i-1} \pmod{q}$$

where $v_0 = 0$, and $r_1, ..., r_{i-1}$ and v_{i-1} are as determined in previous iterations. The prover is supposed to supply the univariant polynomial p_i

$$p_i(z) \overset{\text{def}}{=} \sum_{x_{i+1}=0,1} \cdots \sum_{x_n=0,1} \phi(r_1, ..., r_{i-1}, z, x_{i+1}, ..., x_n) \bmod q$$

Denote by p_i' the actual polynomial sent by the verifier (i.e., the honest prover sets $p_i' = p_i$). Then, the verifier first checks if $p_i'(0) + p_i'(1) \equiv v_{i-1} \pmod{q}$, and next uniformly selects $r_i \in GF(q)$ and sends it to the prover. The claim to be proven in the next iteration is

$$\sum_{x_{i+1}=0,1} \cdots \sum_{x_n=0,1} \phi(r_1, ..., r_{i-1}, r_i, x_{i+1}, ..., x_n) \equiv v_i \pmod{q}$$

where $v_i \overset{\text{def}}{=} p_i'(r_i) \bmod q$.

Completeness of the above: When the claim holds, the prover has no problem supplying the correct polynomials, and this will lead the verifier to always accept.

Soundness of the above: It suffices to bound the probability that, for a particular iteration, the initial claim is false whereas the ending claim is correct. Both claims refer to the current summation expression being equal to the current value, where 'current' means either at the beginning of the iteration or at its end. Let $T(\cdot)$ be the actual polynomial representing the expression when stripping the current variable, and let $p(\cdot)$ be any potential answer by the prover. We may assume that $p(0) + p(1) \equiv v$ and that p is of low-degree (as otherwise the verifier will reject). Using our hypothesis (that the initial claim is false), we know that $T(0) + T(1) \not\equiv v$. Thus, p and T are different low-degree polynomials and so they may agree on very few points (if at all). In case the verifier instantiation (i.e., its choice of random r) does not happen to be one of these few points, the ending claim is false too.

Interactive Proofs for PSPACE. Recall that PSPACE languages can be expressed by Quantified Boolean Formulae. The number of quantifiers is polynomial in the input, but there are both existential and universal quantifiers, and furthermore these quantifiers may alternate. Considering the arithmetization of these formulae, we face two problems: Firstly, the value of the formulae is only bounded by a double exponential function (in the length of the input), and secondly when stripping out summations, the expression may be a polynomial of high degree (due to the universal quantifiers which are replaced by products). The first problem is easy to deal with by using the Chinese Reminder Theorem (i.e., if two integers in $[0, M]$ are different then they must be different modulo most of the primes up-to poly$(\log M)$). The second problem is resolved by "refreshing" variables after each universal quantifier: That is, let $\phi(x_1, ..., x_s, y, x_{s+1}, ..., x_{s+t})$ be a quantifier-free boolean formula with free Boolean variables $x_1, , ..., x_{s+t}, y$, and let $Q_1, ..., Q_{s+t}$ be an arbitrary sequence of quantifiers. Then, we replace the formula

$$Q_1 x_1 \cdots Q_s x_s \forall y Q_{s+1} x_{s+1} \cdots Q_{s+t} x_{s+t} \phi(x_1, ..., x_s, y, x_{s+1}, ..., x_{s+t})$$

by the formula

$$Q_1 x_1 \cdots Q_s x_s \forall y \quad [\quad \exists x'_1 \cdots \exists x'_s \wedge_{i=1}^{s} (x'_i = x_i)$$
$$\wedge Q_{s+1} x_{s+1} \cdots Q_{s+t} x_{s+t} \phi(x'_1, ..., x'_s, y, x_{s+1}, ..., x_{s+t})]$$

This process of refreshing variables is applied from left to right on the entire sequence of quantifiers. Thus, in the resulting formula, no variable quantified to the left of two universal quantifiers may appear on their right. It follows that when arithmetizing and stripping summations (or products) from the resulting quantified Boolean formula, until we get to the very last product, the corresponding univariant polynomial is of constant degree. (The degree of the univariant polynomial obtained when stripping the last product and the rest of the summations is bounded by the number of the original clauses.)

IP in PSPACE: One shows that for every interactive proof system, there exists an optimal prover strategy, and furthermore that this strategy can be computed in polynomial-space (and consequently $\mathcal{IP} \subseteq \mathcal{PSPACE}$). This claim follows by looking at the tree of all possible executions, and observing that the "value" of each node in this tree can be computed recursively in polynomial-space (see related Appendix C.1).

2.2.4 The Interactive Proof System Hierarchy

Concerning the finer structure of the IP-hierarchy, the following is known:

- A "linear speed-up" [27]: For every integer function, f, so that $f(n) \geq 2$ for all n, the class $\mathcal{IP}(O(f(\cdot)))$ collapses to the class $\mathcal{IP}(f(\cdot))$.
  In particular, $\mathcal{IP}(O(1))$ collapses to $\mathcal{IP}(2)$.
- The class $\mathcal{IP}(2)$ contains sets not known to be in $\mathcal{NP}$, e.g., Graph Non-Isomorphism (see above) [185].
- The class $\mathcal{IP}(2)$ is contained in $\mathcal{NP}/\text{poly}$ (i.e., nonuniform-NP), analogously to the containment $\mathcal{BPP} \subseteq \mathcal{P}/\text{poly}$ [330].
- If $\text{co}\mathcal{NP} \subseteq \mathcal{IP}(2)$ then the Polynomial-Time Hierarchy collapses [73].

It is conjectured that $\text{co}\mathcal{NP}$ is *not* contained in $\mathcal{IP}(2)$, and consequently that interactive proofs with an unbounded number of message exchanges are more powerful than interactive proofs in which only a bounded (i.e., constant) number of messages are exchanged. The class $\mathcal{IP}(1)$ (also denoted $\mathcal{MA}$) seems to be *the* "real" randomized (and yet non-interactive) version of $\mathcal{NP}$: Here the prover supplies a candidate (polynomial-size) "proof", and the verifier assesses its validity probabilistically (rather than deterministically). We note that certain derandomization results regarding $\mathcal{BPP}$ imply that $\mathcal{IP}(1) = \mathcal{NP}$. Specifically, if any *promise problem* [132] solvable in probabilistic polynomial-time is solvable in deterministic polynomial-time then $\mathcal{IP}(1) = \mathcal{NP}$. Fur further discussion see [195].

Variants. As mentioned above, the IP-hierarchy (i.e., $\mathcal{IP}(\cdot)$) equals an analogous hierarchy, denoted $\mathcal{AM}(\cdot)$, in which the verifier is restricted to send the outcome of any coin it tosses [203]. The latter restricted proof systems are called *Arthur-Merlin games* or *public-coin interactive proofs*. In addition, the IP-hierarchy equals an analogous two-sided error hierarchy [158]. In the latter proof systems the completeness condition is relaxed so that the verifier is required to accept each $x \in L$ with probability at least $\frac{2}{3}$. In both cases, we mean that for every integer function f with $f(n) \geq 1$ (for all n's), the f-level of the alternative hierarchy coincides with the f-level of the basic hierarchy (i.e., $\mathcal{IP}(f)$).[5] Thus, the constant levels of all hierarchies coincide with $\mathcal{AM} \overset{\text{def}}{=} \mathcal{AM}(2)$.

---

[5] For $f \equiv 1$ there is no syntactic difference between Arthur-Merlin games and interactive proof systems. As for one-sided versus two-sided error, the transformation of [158] adds an initial message by the prover, which can be incorporated into the single message sent in a one-message interactive proof system. In gene-

2.2.5 How Powerful Should the Prover Be?

Assume that a set S is in $\mathcal{IP}$. This means that there is a verifier V that can be convinced to accept any input in S but cannot be convinced to accept any input not in S (except with small probability). One may ask how powerful should a prover be so that it can convince the verifier V to accept any input in S. More interestingly, considering all possible verifiers which give rise to interactive proof systems for S, what is the minimum power required from a prover which satisfies the completeness requirement with respect to one of these verifiers?

We stress that, unlike the case of computationally-sound proof systems (see Sec. 2.5), we do not restrict the power of the prover in the soundness condition, but rather consider the minimum complexity of provers meeting the completeness condition. Specifically, we are interested in *relatively efficient* provers which meet the completeness condition. The term 'relatively efficient prover' has been given three different interpretations.

1. A prover is considered *relatively efficient* if, when given an auxiliary input (in addition to the common input in S), it works in (probabilistic) polynomial-time. Specifically, in case $S \in \mathcal{NP}$, the auxiliary input maybe an NP-proof that the common input is in the set[6].

   This interpretation is adequate and in fact crucial for applications in which such an auxiliary input is available to the otherwise-polynomial-time parties. Typically, such auxiliary input is available in cryptographic applications in which parties wish to prove in (zero-knowledge) that they have conducted some computation correctly. In these cases the NP-proof is just the transcript of the computation by which the claimed result has been generated, and thus the auxiliary input is available to the proving party. See [185].

2. A prover is considered *relatively efficient* if it can be implemented by a probabilistic polynomial-time oracle machine with oracle access to the set S itself. (Note that the prover in Construction 2.3 has this property.) This interpretation generalizes the notion of self-reducibility of NP-sets. (By self-reducibility of an NP-set we mean that the search problem of finding an NP-witness is polynomial-time reducible to deciding membership in the set.) See [41].

3. A prover is considered *relatively efficient* if it can be implemented by a probabilistic machine which runs in time which is polynomial in the

---

ral, for $f \geq 2$, the transformations of [158] and [203] may add 1 or 2 messages, respectively, but this effect may be removed using the linear speed-up result of [27] (mentioned above). As for the 0th level it is not interactive in any sense; it is syntactically equal to $\mathcal{BPP}$ or to $\text{co}\mathcal{RP}$ (depending on whether we consider two-sided or one-sided error).

[6] Still, even in this case the interactive proof need not consist of the prover sending the auxiliary input to the verifier; e.g., an alternative procedure may allow the prover to be zero-knowledge (see Construction 2.7).

deterministic complexity of the set. This interpretation relates the difficulty of convincing a "lazy verifier" to the complexity of finding the truth alone.

Hence, in contrast to the first interpretation which is adequate in settings where assertions are generated along with their NP-proofs, the current interpretation is adequate in settings in which the prover is given only the assertion and has to find a proof to it by itself (before trying to convince a lazy verifier of its validity). See [267].

2.3 Zero-Knowledge Proof Systems

Zero-knowledge proofs, introduced in [199], are central to cryptography. Furthermore, zero-knowledge proofs are very intriguing from a conceptual point of view, since they exhibit an extreme contrast between being convinced of the validity of a statement and learning anything in addition while receiving such a convincing proof. Namely, zero-knowledge proofs have the remarkable property of being both convincing while yielding nothing to the verifier, beyond the fact that the statement is valid. Formally, the fact that "nothing is gained by the interaction" is captured by stating that whatever the verifier can efficiently compute after interacting with a zero-knowledge prover, can be efficiently computed from the assertion itself, without interacting with anyone.

2.3.1 A Sample Definition

Zero-knowledge is a property of some interactive proof systems, or more accurately of some specified prover strategies. The formulation of the zero-knowledge condition considers two ensembles of probability distributions, each ensemble associates a probability distribution to each valid assertion. The first ensemble represents the output distribution of the verifier after interacting with the specified prover strategy P, where the verifier is not necessarily employing the specified strategy (i.e., V) – but rather any efficient strategy. The second ensemble represents the output distribution of some probabilistic polynomial-time algorithm (which does not interact with anyone). The basic paradigm of zero-knowledge asserts that for every ensemble of the first type there exist a "similar" ensemble of the second type. The specific variants differ by the interpretation given to 'similarity'. The most strict interpretation, leading to *perfect zero-knowledge*, is that similarity means equality. Namely,

Definition 2.5 (perfect zero-knowledge, simplified[7] [199]): *A prover strategy, P, is said to be* perfect zero-knowledge *over a set S if for every pro-*

---

[7] In the actual definition one either allows M^* to run for *expected* polynomial-time (as done in [199, 185]) or allows M^* to have no output with probability at most

babilistic polynomial-time verifier strategy, V^*, there exists a probabilistic polynomial-time algorithm, M^*, such that

$$(P, V^*)(x) \equiv M^*(x) , \quad \text{for every } x \in S$$

where $(P, V^*)(x)$ is a random variable representing the output of verifier V^* after interacting with the prover P on common input x, and $M^*(x)$ is a random variable representing the output of machine M^* on input x.

A somewhat more relaxed interpretation, leading to *almost-perfect zero-knowledge*, is that similarity means statistical closeness (i.e., negligible difference between the ensembles). The most liberal interpretation, leading to the standard usage of the term zero-knowledge (and sometimes referred to as *computational zero-knowledge*), is that similarity means computational indistinguishability (i.e., failure of any efficient procedure to tell the two ensembles apart). Since the notion of computational indistinguishability is a fundamental one, it is indeed in place to present a definition of it.

Definition 2.6 (computational indistinguishability [198, 351]): *An integer function, f, is called* negligible *if for every positive polynomial p and all sufficiently large n, it holds that $f(n) < \frac{1}{p(n)}$.* (Thus, multiplying a negligible function by any fixed polynomial yields a negligible function.)
Two probability ensembles, $\{A_x\}_{x \in S}$ and $\{B_x\}_{x \in S}$, are indistinguishable *by an algorithm D if*

$$d(n) \overset{\text{def}}{=} \max_{x \in S \cap \{0,1\}^n} \{|\Pr[D(A_x)=1] - \Pr[D(B_x)=1]|\}$$

is a negligible function. The ensembles $\{A_x\}_{x \in S}$ and $\{B_x\}_{x \in S}$ are computationally indistinguishable *if they are indistinguishable by every probabilistic polynomial-time algorithm.*

The definitions presented above are a simplified version of the actual definitions. For example, in order to guarantee that zero-knowledge is preserved under sequential composition it is necessary to slightly augment the definitions. For details see [187].

Knowledge Complexity. Zero-knowledge is the lowest level of a knowledge-complexity hierarchy which quantifies the "knowledge revealed in an interaction" [199]. *Knowledge complexity* may be defined as the minimum number of oracle-queries required in order to (efficiently) simulate an interaction with the prover (cf., [189]). Results linking two different variants of this measure to other complexity measures are given in [1, 297], respectively.

1/2 (as done in [170]). The latter alternative implies the former, but the converse is not known to hold.

2.3.2 The Power of Zero-Knowledge

A simple example, demonstrating the power of zero-knowledge proofs, follows. Specifically, we will present a simple zero-knowledge proof for proving that a graph is 3-colorable[8]. The interactive proof will be described using "boxes" in which information can be hidden and later revealed. Such "boxes" can be implemented using one-way functions (see below).

Construction 2.7 (Zero-knowledge proof of 3-colorability [185]):

- Common Input: *A simple graph $G = (V, E)$.*
- Prover's first step: *Let ψ be a 3-coloring of G. The prover selects a random permutation, π, over $\{1, 2, 3\}$, and sets $\phi(v) \stackrel{\text{def}}{=} \pi(\psi(v))$, for each $v \in V$. Hence, the prover forms a random relabeling of the 3-coloring ψ. The prover sends the verifier a sequence of $|V|$ locked and non-transparent boxes so that the vth box contains the value $\phi(v)$;*
- Verifier's first step: *The verifier uniformly selects an edge $\{u, v\} \in E$, and sends it to the prover;*
- Motivating Remark: *The verifier asks to inspect the colors of vertices u and v;*
- Prover's second step: *The prover sends to the verifier the keys to boxes u and v;*
- Verifier's second step: *The verifier opens boxes u and v, and accepts if and only if they contain two different elements in $\{1, 2, 3\}$;*

The verifier strategy presented above is easily implemented in probabilistic polynomial-time. The same holds with respect to the prover's strategy, provided it is given a 3-coloring of G as auxiliary input. Clearly, if the input graph is 3-colorable then the prover can cause the verifier to accept always. On the other hand, if the input graph is not 3-colorable then any contents put in the boxes must be invalid on at least one edge, and consequently the verifier will reject with probability at least $\frac{1}{|E|}$. Hence, the above game exhibits a non-negligible gap in the accepting probabilities between the case of 3-colorable graphs and the case of non-3-colorable graphs. To increase the gap, the game may be repeated sufficiently many times (of course, using independent coin tosses in each repetition). The zero-knowledge property follows easily, in this abstract setting, since one can simulate the real interaction by placing a random pair of different colors in the boxes indicated by the verifier. This indeed demonstrates that the verifier learns nothing from the interaction (since it expects to see a random pair of different colors and indeed this is what it sees). We stress that this simple argument is not possible in the digital implementation since the boxes are not totally unaffected by their contents (but are rather effected, yet in an indistinguishable manner). Instead, we simulate

---

[8] A graph $G = (V, E)$ is said to be *3-colorable* if there exists a function $\pi : V \mapsto \{1, 2, 3\}$ so that $\pi(v) \neq \pi(u)$ for every $\{u, v\} \in E$. Such a function, π, is called a *3-coloring* of the graph.

the interaction as follows. We first guess (at random) which pair of boxes the verifier would ask to open, and place a random pair of distinct colors in these boxes (and garbage in the rest). We hand all boxes to the verifier. In case the verifier asks for the chosen pair (i.e., the one we guessed), we can complete the simulation. Otherwise, we try again (with a new random guess). Thus, it suffices to use boxes which hide their contents quite well (rather than being perfectly opaque). Such boxes can be implemented digitally.

Digital implementation. We implement the "boxes" (used above) by using an adequately defined "commitment scheme". Loosely speaking, such a scheme is a two phase game between a sender and a receiver so that after the first phase the sender is "committed" to a value and yet, at this stage, it is infeasible for the receiver to find out the committed value. The committed value will be revealed to the receiver in the second phase and it is guaranteed that the sender cannot reveal a value other than the one committed. Such commitment schemes can be implemented assuming the existence of one-way functions (i.e., loosely speaking, functions that are easy to compute but hard to invert, such as the multiplication of two large primes) [272, 211].

Using the fact that 3-colorability is NP-complete, one gets zero-knowledge proofs for any NP-set.

Theorem 2.8 (The ZK Theorem [185]): *Assuming the existence of one-way functions, any NP-proof can be efficiently transformed into a* (computational) *zero-knowledge interactive proof.*

The hypothesis (regarding the existence of one-way functions) in the above theorem seems unavoidable – the existence of zero-knowledge proofs for "hard on the average" problems implies the existence of one-way functions (and, likewise, the existence of zero-knowledge proofs for sets outside $\mathcal{BPP}$ implies the existence of "auxiliary-input one-way functions") [292]. Theorem 2.8 has a dramatic effect on the design of cryptographic protocols (cf., [185, 186]). In a different vein and for the sake of elegancy, we mention that, using further ideas and under the same assumption, any interactive proof can be efficiently transformed into a zero-knowledge one [222, 57]. Thus,

Theorem 2.9 (The ultimate ZK Theorem [222, 57]): *Assuming the existence of one-way functions, $\mathcal{IP} = \mathcal{CZK}$, where $\mathcal{CZK}$ is the class of sets having* (computational) *zero-knowledge proof systems.*

Perfect and Statistical Zero-Knowledge. The above results may be contrasted with the results regarding the complexity of *almost-perfect* (a.k.a statistical) zero-knowledge proof systems: Almost-perfect zero-knowledge proof systems exist only for sets in $\mathcal{IP}(2) \cap \text{co}\mathcal{IP}(2)$ [153, 2], and thus are unlikely to exist for all NP-sets. On the other hand, the class Statistical Zero-Knowledge is known to contain some hard problems (cf., discussion in [192]), and turns out to have interesting complexity theoretic properties (cf., [290, 316, 192]).

2.3.3 The Role of Randomness

Again, randomness is essential to all the above mentioned (positive) results. Namely, if either the verifier or the prover is required to be deterministic then only BPP-sets can be proven in a zero-knowledge manner [187]. How. ver, BPP-sets have trivial zero-knowledge proofs in which the prover sends nothing and the verifier just test the validity of the assertion by itself.[9] Thus, randomness is essential to the usefulness of zero-knowledge proofs.

2.4 Probabilistically Checkable Proof Systems

When viewed in terms of an interactive proof system, the probabilistically checkable proof setting consists of a prover which is memoryless. Namely, one can think of the prover as being an oracle and of the messages sent to it as being queries. A more appealing interpretation is to view the probabilistically checkable proof setting as an alternative way of generalizing $\mathcal{NP}$. Instead of receiving the entire proof and conducting a deterministic polynomial-time computation (as in the case of $\mathcal{NP}$), the verifier may toss coins and query the proof only at location of its choice. Potentially, this allows the verifier to utilize very long proofs (i.e., of super-polynomial length) or alternatively examine very few bits of an NP-proof.

2.4.1 Definition

Loosely speaking, a probabilistically checkable proof system consists of a probabilistic polynomial-time verifier having access to an oracle which re-presents a proof in redundant form. Typically, the verifier accesses only few of the oracle bits, and these bit positions are determined by the outcome of the verifier's coin tosses. Again, it is required that if the assertion holds then the verifier always accepts (i.e., when given access to an adequate oracle); whereas, if the assertion is false then the verifier must reject with probability at least $\frac{1}{2}$, no matter which oracle is used. The basic definition of the PCP setting is given in Item (1) below. Yet, the complexity measures introduced in Item (2) are of key importance for the subsequent discussions, and should not be ignored.

Definition 2.10 (Probabilistic Checkable Proofs – PCP):

1. *A* probabilistic checkable proof system (pcp) *for a set S is a probabilistic polynomial-time oracle machine* (called verifier)*, denoted V, satisfying*

---

[9] Actually, this is slightly inaccurate since the resulting "interactive proof" may have two-sided error, whereas we have required interactive proofs to have only one-sided error. Yet, since the error can be made negligible by successive repetitions this issue is insignificant.

 – Completeness: *For every $x \in S$ there exists an oracle π_x so that V, on input x and access to oracle π_x, always accepts x.*
 – Soundness: *For every $x \notin S$ and every oracle π, machine V, on input x and access to oracle π, rejects x with probability at least $\frac{1}{2}$.*

2. *Let r and q be integer functions. The complexity class $\mathcal{PCP}(r(\cdot), q(\cdot))$ consists of sets having a probabilistic checkable proof system in which the verifier, on any input of length n, makes at most $r(n)$ coin tosses and at most $q(n)$ oracle queries. We stress that here, as usual in complexity theory, the oracle answers are always binary (i.e., either 0 or 1).*

   For sets of integer functions, R and Q, we let $\mathcal{PCP}(R, Q)$ equal $\bigcup_{r \in R, q \in Q} \mathcal{PCP}(r(\cdot), q(\cdot))$.

The above model was suggested in [154] and shown related to a multi-prover model introduced previously in [58]. The fine complexity measures were introduced and motivated in [139], and further advocated in [20]. A related model was presented in [25], stressing the applicability to program checking.

We stress that the oracle π_x in a pcp system constitutes a proof in the standard mathematical sense. (Jumping ahead, the oracles in pcp systems characterizing $\mathcal{NP}$ have the property of being NP proofs themselves.) Yet, this oracle has the extra property of enabling a lazy verifier, to toss coins, take its chances and "assess" the validity of the proof without reading all of it (but rather by reading a tiny portion of it).

2.4.2 The Power of Probabilistically Checkable Proofs

Clearly, $\mathcal{PCP}(\text{poly}, 0)$ equals $\text{co}\mathcal{RP}$, whereas $\mathcal{PCP}(0, \text{poly})$ equals $\mathcal{NP}$. It is easy to prove an upper bound on the non-deterministic time complexity of sets in the PCP hierarchy. In particular,

Proposition 2.11 *$\mathcal{PCP}(\log, \text{poly})$ is contained in $\mathcal{NP}$.*

The above follows by observing that PCP systems of logarithmic randomness only utilize a polynomial (in the input length) portion of the oracle. This observation also explains much of the appeal of such proof systems – the oracle in such PCP systems constitutes an NP-proof with extra properties; we refer to the ability to evaluate the validity of this proof by reading a small portion of it. Thus, any result of the form

$$\mathcal{NP} \subseteq \mathcal{PCP}(\log, q(\cdot)) \tag{2.1}$$

where q is any fixed polynomial would have been interesting (as it would apply also to NP-sets having witnesses of length exceeding $q(n)$), and the smaller q – the better. Interestingly, the polynomial q can be made a constant, and this fact – known as the PCP Theorem – has very important consequences. The PCP Theorem is a culmination of a sequence of great works [24, 25, 139,

20, 19],[10] each establishing meaningful and increasingly stronger versions of
Eq. (2.1). An overview of the proof is given below.

Theorem 2.12 (The PCP Theorem [19]):

$$\mathcal{NP} \text{ is contained in } \mathcal{PCP}(\log, O(1)).$$

Thus, probabilistically checkable proofs in which the verifier tosses only logarithmically many coins and makes only a constant number of queries exist
for every set in the complexity class $\mathcal{NP}$. Furthermore, the proof of Theorem 2.12 is constructive in the sense that it allows to efficiently transform
any NP-witness (for an instance of a set in $\mathcal{NP}$) into an oracle which makes
the PCP verifier always accept. Thus, NP-proofs can be transformed into
NP-proofs which offer a trade-off between the portion of the proof being read
and the confidence it offers. Specifically, for every $\epsilon > 0$, if the verifier is
willing to tolerate an error probability of ϵ then it suffices to let it examine
$O(\log(1/\epsilon))$ bits of the (transformed) NP-proof. These bit locations need to
be selected at random.

Combining Theorem 2.12 with Proposition 2.11 we obtain the following characterization of $\mathcal{NP}$.

Corollary 2.13 (The PCP characterization of NP):
$\mathcal{NP} = \mathcal{PCP}(\log, O(1)).$

Overview of the Proof of Theorem 2.12

The proof of the PCP Theorem (Theorem 2.12) is one of the most complicated
proofs in the Theory of Computation. Its main ingredients are:

1. A $\mathcal{PCP}(\log, \text{poly}(\log))$ proof system for $\mathcal{NP}$. Furthermore, this proof
   system has additional properties which enable proof composition as in
   item (3) below.
2. A $\mathcal{PCP}(\text{poly}, O(1))$ proof system for $\mathcal{NP}$. This proof system also has
   additional properties enabling proof composition as in item (3).
3. The proof composition paradigm: In general this paradigm allows to com-
   pose two proof systems so that the "inner" one is used to probabilistically
   verify the acceptance criteria of the "outer" verifier. The aim is to con-
   duct this verification using fewer queries than the total query complexity
   of the "outer" proof system. This is done by encoding the supposed answers of the "outer" system using an appropriate error correcting code.
   Thus, the "inner" verifier should be able to verify claims made with respect to an encoded input, presented by an input oracle, using much

---

[10] See Section 2.6.2 for an account of the developments leading to Theorem 2.12.
The constant (number of queries) in Theorem 2.12 has been subsequently improved, and is currently 5; cf., [40, 210, 205].

fewer queries than the length of the input. Actually, the "inner" verifier should be able to process inputs presented by several such oracles, and the "outer" verifier should operate by making at most a corresponding number of queries (possibly to a multi-valued oracle).

Suppose we are given a $\mathcal{PCP}(r(\cdot), O(\ell(\cdot)))$ system for $\mathcal{NP}$ in which a constant number of queries are made (non-adaptively) to an 2^ℓ-valued oracle, and the verifier's decision regarding the answers may be implemented by a poly(ℓ)-size circuit. Further suppose that we are given a $\mathcal{PCP}(r'(\cdot), q(\cdot))$-like system for $\mathcal{P}$ in which the input is given in encoded form via an additional oracle so that the system accepts input-oracles which encode inputs in the language and reject any input-oracle which is "far" from the encoding of any input in the language. In this latter system access to the input-oracle is accounted in the query complexity. Furthermore, suppose that the latter system may handle inputs which result from concatenation of a constant number of sub-inputs, each encoded in a separate sub-input oracle.

Then, $\mathcal{NP} \subseteq \mathcal{PCP}(2(r(\cdot) + r'(s(\cdot))), 2q(s(\cdot)))$, where $s(n) \overset{\text{def}}{=} \text{poly}(\ell(n))$. [The extra factor of 2 is an artifact of the need to amplify each of the two proof systems so that the total error probability sums up to at most $1/2$.]

In particular, the proof system of item (1) is composed with itself [using $r = r' = \log$, $\ell = q = \text{poly}(\log)$, and $s(n) = \text{poly}(\log(n))$] yielding a $\mathcal{PCP}(\log, \text{poly}(\log\log))$ system for $\mathcal{NP}$, which is then composed with the system of item (2) [using $r = \log$, $\ell = \text{poly}(\log\log)$, $r' = \text{poly}$, $q = O(1)$, and $s(n) = \text{poly}(\log\log(n))$] yielding the desired $\mathcal{PCP}(\log, O(1))$ system for $\mathcal{NP}$.

The $\mathcal{PCP}(\log, \text{poly}(\log))$ system for $\mathcal{NP}$: We start with a different arithmetization of CNF formulae (i.e., other than the one used for constructing an interactive proof for co$\mathcal{NP}$). Logarithmically many variables are used to represent (in binary) the names of variables and of clauses in the input formula, and an oracle from variables to Boolean values is supposed to represent a satisfying assignment. An arithmetic expression involving a logarithmic number of summations is used to represent the value of the formula under the truth assignment represented by the oracle. This expression is a low-degree polynomial in the new variables and has a cubic dependency on the assignment-oracle. Small-biased probability spaces are used to generate a polynomial number of such expressions so that if the formula is satisfiable then all these expressions evaluate to zero, and otherwise at most half of them evaluate to zero. Using a summation test (as in the interactive proof for co$\mathcal{NP}$) and a low-degree test, this yields a $\mathcal{PCP}(t(\cdot), t(\cdot))$ system for $\mathcal{NP}$, where $t(n) \overset{\text{def}}{=} O(\log(n) \cdot \log\log(n))$. [We use a finite field of poly($\log(n)$) elements, and so we need $(\log n) \cdot O(\log\log n)$ random bits for the summation and low-degree tests.] To obtain the desired pcp system, one uses $\frac{O(\log n)}{\log\log n}$-long sequences over $\{1, ..., \log n\}$ to represent variable/clause names (rather than logarithmically-long binary sequences). [We can still use a finite field of poly$(\log(n))$ elements, and so we need only $\frac{O(\log n)}{\log \log n} \cdot O(\log \log n)$ random bits for the summation and low-degree tests.] All this is relatively easy compared to what is needed in order to transform the pcp system so that only a constant number of queries are made to a (multi-valued) oracle. This is obtained via a (randomness-efficient) "parallelization" of pcp systems, which in turn depends heavily on efficient low-degree tests. (Indeed, this "parallelization" is the most technically complex part of the entire proof of the PCP Theorem.)

The $\mathcal{PCP}(\text{poly}, O(1))$ system for $\mathcal{NP}$: It suffices to prove the satisfiability of a systems of quadratic equations over GF(2) (as this problem too is NP-complete). The oracle is supposed to hold the values of all quadratic expressions under a satisfying assignment to the (say n) variables. We distinguish two tables in the oracle: One corresponding to the (2^n) linear expressions and the other to the $(2^{n^2}$ pure) bilinear expressions. Each table is tested for self-consistency (via a linearity test), and the two tables are tested to be consistent with each other (via a matrix-equality test which utilizes "self-correction"). Each of these tests utilizes a constant number of Boolean queries, and randomness which is logarithmic in the size of the corresponding table (and thus poly(n)).

2.4.3 PCP and Approximation

The characterization of $\mathcal{NP}$ in terms of probabilistically checkable proofs plays a central role in recent developments concerning the difficulty of approximation problems (cf., [139, 19, 256, 40] and [209, 210]). To demonstrate this relationship, we first note that Theorem 2.12 can be rephrased without mentioning the class $\mathcal{PCP}$ altogether. Instead, a new type of polynomial-time reductions, which we call *amplifying*, emerges.

Theorem 2.14 (Theorem 2.12 — Rephrased): *There exists a constant $\epsilon > 0$, and a polynomial-time computable function f, mapping the set of 3CNF formulae[11] to itself so that*

— *As usual, f maps satisfiable 3CNF formulae to satisfiable 3CNF formulae; and*
— *f maps non-satisfiable 3CNF formulae to* (non-satisfiable) *3CNF formulae for which every truth assignment satisfies at most a $1 - \epsilon$ fraction of the clauses.*

The function f is called an amplifying reduction.

---

[11] A 3CNF formula is a Boolean formula consisting of a conjunction of clauses, where each clause is a disjunction of upto 3 literals. (A literal is variable or its negation.).

Proof Sketch (Thm. 2.12 ⇒ Thm. 2.14): We start by considering a pcp system for 3SAT, and use the fact that the pcp system given by the proof of Theorem 2.12 is non-adaptive (i.e., the queries are determined as a function of the input and the random-tape – and do not depend on answers to previous queries).[12] Next, we associate the bits of the oracle (of this pcp system) with Boolean variables, and introduce a (constant size) Boolean formula for each possible outcome of the sequence of $O(\log n)$ coin tosses, describing whether the verifier would have accepted given this outcome. (For each input and each fix outcome of the coin tosses, the verifier's decision depends only on a constant number or oracle bits.) Finally, using auxiliary variables, we convert each of these formulae into a 3CNF formula and obtain (as the output of the reduction) the conjunction of all these polynomially-many clauses.   □

It is also easy to see that Theorem 2.14 implies Theorem 2.12: Given a reduction as in Theorem 2.14, we construct a pcp system for 3SAT by letting the verifier select a clause uniformly among the clauses of the reduced formula, and make three queries corresponding to the three variables in it. This yields a proof system with soundness error bounded by $1 - \epsilon$. Theorem 2.12 is obtained by reducing the error probability, using $O(1/\epsilon)$ successive applications of the proof system.

As an immediate corollary to the formulation of Theorem 2.14 one concludes that it is NP-Hard to distinguish satisfiable 3CNF formulae from 3CNF formulae for which no truth assignment satisfies at least a $1 - \epsilon$ fraction of the clauses (as otherwise, using the reduction, one may decide membership in 3SAT). In general, probabilistic checkable proof systems for $\mathcal{NP}$ yield strong non-approximability results for various classical optimization problems. In particular, quite *tight* non-approximability results have been shown for Max-Clique (cf., [209]), Chromatic Number (cf., [141]), Set Cover (cf., [134]), and Max3SAT (cf., [210] and algorithm in [228]). For further details the reader is referred to [18] (alas this survey does not contain the most recent results).

2.4.4 More on PCP Itself

We start by discussing variants of the PCP characterization of NP, and next turn to PCPs having expressing power beyond NP.

More on the PCP characterization of NP. Interestingly, the two complexity measures in the PCP-characterization of $\mathcal{NP}$ can be traded off, so that at the extremes we get $\mathcal{NP} = \mathcal{PCP}(\log, O(1))$ and $\mathcal{NP} = \mathcal{PCP}(0, \texttt{poly})$, respectively.

Proposition 2.15 : *There exist constants $\alpha, \beta > 0$ such that for every integer function $l(\cdot)$, so that $0 \leq l(n) \leq \alpha \log_2 n$,*

---

[12] Actually, it is not essential to use this fact, since one can easily convert any adaptive system into a non-adaptive one while incurring an exponential blowup in the query complexity (which in our case is a constant).

$$\mathcal{NP} = \mathcal{PCP}(r(\cdot), q(\cdot)),$$

where $r(n) = \alpha \cdot \log_2 n - l(n)$ and $q(n) = \beta \cdot 2^{l(n)}$.

Proof Idea: Starting with Theorem 2.12, one tries all possibilities for the $l(n)$-long prefix of the random tape of the verifier.   □

The above simple observation is but the tip of an iceberg. In the years which have passed since the establishment of Theorem 2.12 many far more interesting and technically involved facts regarding the PCP characterization of NP were discovered. Following is a brief summary of the various (still active) research directions.

− *The length of PCPs:* By definition, the number of possible different oracle queries in a $\mathcal{PCP}(\log, \log)$ system is polynomial (in the length of the input). Actually, in the PCP systems of Theorem 2.12 these queries refer only to a polynomially long prefix of the oracle, and so we may say that the length of these PCPs for $\mathcal{NP}$ is polynomial. It is known that the length of PCPs for $\mathcal{NP}$ can be made nearly-linear if one allows poly-logarithmically many queries [301]. Further understanding of the trade-off is indeed due.
− *The number of queries in PCPs:* Theorem 2.12 asserts that a constant number of queries suffice for PCPs with logarithmic randomness and soundness error 1/2 (for NP). It is currently known that this constant is at most 5 (whereas with 3 queries one may get arbitrary close to error 1/2) [205]. Allowing an arbitrary small constant error in the completeness condition, 3 queries are sufficient [210] (and necessary, unless $\mathcal{P} = \mathcal{NP}$). The obvious trade-off between the number of queries and the soundness error gives rise to the robust notion of *amortized query complexity* defined as the ratio of the number of queries and (minus) the logarithm (to based 2) of the soundness error. It is known that PCPs of logarithmic randomness and amortized query complexity 1 exist only for sets in $\mathcal{P}$ [40]. On the other hand, PCPs of logarithmic randomness and amortized query complexity $2.5 + \epsilon$ exists for all $\mathcal{NP}$ and any $\epsilon > 0$ (see [205] presenting a 5-query system of error $0.25 + \epsilon$). The gap in case one allows arbitrary small constant error probability in the completeness condition is less wide − here the bounds are 1 (again) versus $1.5 + \epsilon$, for any $\epsilon > 0$ [335].
− *The free-bit complexity:* The motivation to this notion came from the PCP–MaxClique connection, but we find it intriguing for its own sake. Loosely speaking, here one distinguishes queries for which the verifier compares the answer with a value determined by previously obtained answers, from queries in which the verifier only records the answer for future usage [140]. The latter queries are called *free* (as the "acceptable answers" to them are not determined). The *amortized free-bit complexity* is define analogously [54]. Interestingly, $\mathcal{NP}$ has PCPs with logarithmic randomness and amortized free-bit complexity less than any positive constant (cf., Håstad [209]).

- *Adaptive versus non-adaptive*: A PCP verifier is called *non-adaptive* if its queries are determined solely based on its input and the outcome of its coin tosses. (A general verifier, called adaptive, may determine its queries also based on previously received oracle answers.) Recall that the PCP Characterization of NP (i.e., Theorem 2.12) is obtained using a non-adaptive verifier; however, it turns out that adaptive verifier *are* more powerful than non-adaptive ones (in terms of quantitative results): Specifically, for every $\epsilon > 0$ and logarithmic randomness, (adaptive) 3-query PCPs with soundness error $0.5 + \epsilon$ exist for $\mathcal{NP}$ [205], whereas non-adaptive verifiers making 3 queries and having soundness error 5/8 exist only for $\mathcal{P}$ [357].
- *Non-binary queries*: Our definition of PCP allows only binary queries. Certainly, non-binary queries can always be coded as binary ones, but the converse is not necessarily valid, in particular in adversarial settings. Note that the soundness condition constitutes an implicit adversarial setting, where a bad proof may be thought of as being selected by an adversary. Thus, when several binary queries are packed into one non-binary query, the adversary need not respect the packing (i.e., it may answer inconsistently on the same binary query depending on the other queries packed with it). For this reason, "parallel repetition" is highly non-trivial in the PCP (as well as the MIP) setting; see [309]. Still, using adequate "consistency tests" one may construct PCP systems for $\mathcal{NP}$ using logarithmic randomness, a constant number of queries and soundness error exponential in the length of the answers (cf., [310] as well as [21]). (Currently, this is known only for sub-logarithmic answer lengths.) We comment that 2 non-binary queries are known to be less powerful (in terms of quantitative results) than an equivalent number of binary queries [335].

PCP with super-logarithmic randomness. The above text has focused on the important case where the verifier tosses logarithmically many coins, and hence the "effective proof length" is polynomial. Here we shortly mention that the above main results scale up as follows.

Proposition 2.16 (Proposition 2.11 – Generalized): *For every integer function $r(\cdot)$, the class $\mathcal{PCP}(r(\cdot), \texttt{poly})$ is contained in* $\mathrm{Ntime}(2^{O(r(\cdot)+\log(\cdot))})$.

We comment that $\mathcal{PCP}(o(\log), o(\log)) = \mathcal{P}$ (this follows from the FGLSS-reduction [139] to MaxClique, cf., [20]).

Theorem 2.17 (Theorem 2.12 – Generalized): *Let $t(\cdot)$ be an integer function so that $n < t(n) < 2^{\mathrm{poly}(n)}$, for all n's. Then, the class $\mathrm{Ntime}(t(\cdot))$ is contained in the class $\mathcal{PCP}(O(\log t(\cdot)), O(1))$.*

2.4.5 The Role of Randomness

No trade-off, between the number of bits examined and the confidence, is possible if one requires the verifier to be deterministic. In particular, $\mathcal{PCP}(0, q(\cdot))$

contains only sets that are decidable by a deterministic algorithms of running time $2^{q(n)} \cdot \text{poly}(n)$. It follows that $\mathcal{PCP}(0, \log) = \mathcal{P}$. Furthermore, since it is unlikely that all NP-sets can be decided by (deterministic) algorithms of running time, say, $2^n \cdot \text{poly}(n)$, it follows that $\mathcal{PCP}(0, n)$ is unlikely to contain $\mathcal{NP}$.

2.5 Other Probabilistic Proof Systems

In this section, we shortly review some variants on the basic model of inter-active proofs. These variants include models in which the prover is restricted in its choice of strategy, a model in which the prover-verifier interaction is restricted, and a model in which one proves "knowledge" of facts rather than their *validity*.

2.5.1 Restricting the Prover's Strategy

We stress that the restrictions discussed here refer to the strategies employed by the prover both in case it tries to prove valid assertions (i.e., the completeness condition) and in case it tries to fool the verifier to believe false statements (i.e., the soundness condition). Thus, the validity of the verifier decision (concerning false statements) depends on whether this restriction (concerning "cheating" prover strategies) really holds. The reason to consider these restricted models is that they enable to achieve results which are not possible in the general model of interactive proofs (cf., [58, 77, 232, 265]). We consider restrictions of two types – computational and physical. We start with the latter.

Multi-Prover Interactive Proof Systems (MIP): In the so-called *multi-prover interactive proof* model, denoted MIP (cf., [58]), the prover is split into several (say, two) entities and the restriction (or assumption) is that these entities cannot interact with each other. Actually, the formulation allows them to coordinate their strategies prior to interacting with the verifier[13] but it is crucial that they don't exchange messages among themselves while interacting with the verifier. The multi-prover model is reminiscent of the common police procedure of isolating collaborating suspects and interrogating each of them separately. A typical application in which the two-prover model may be assumed is an ATM that verifies the validity of a pair of smart-cards inserted in two isolated slots of the ATM. The advantage in using such a split system is that it enables the presentation of (perfect) zero-knowledge proof systems for any set in $\mathcal{NP}$, using no intractability assumptions [58]. Furthermore, these proofs can be made very efficient in terms of communication complexity [126]. Interestingly, the multi-prover model is related to the PCP model [154]; yet the relationship is not straightforward (cf., [42, 336]). (In fact, the multi-prover formulation was the one presented first.)

---

[13] This is implicit in the universal quantifier used in the soundness condition.

Computationally-Sound Proof Systems (arguments and CS-Proofs): We now turn to computational restrictions. Since the effect of this restriction is more noticeable in the soundness condition, we refer to these proof systems as being *computationally-sound*. Two variants have been suggested. In *argument* systems [77], the prover strategy is restricted to be probabilistic polynomial-time with auxiliary input (analogously to item (1) in Sec. 2.2.5). In *CS-proofs* [267], the prover strategy is restricted to be probabilistic and run in time polynomial in the time required to validate the assertion (analogously to item (3) in Sec. 2.2.5). Interestingly, computationally-sound interactive proofs can be much more communication-efficient than (regular) interactive proofs (cf. [232, 267, 176]). Details follow.

Argument Systems. The definition of an argument system is derived from the definition of an interactive proof system by modifying the completeness and soundness conditions as follows.

- *Completeness*: The prover P runs in time polynomial in the length of the common input. For every $x \in S$, there exists an auxiliary input (for the prover), w_x, so that the verifier V always accepts after interacting with $P(w_x)$ on common input x.
- *Soundness*: For every probabilistic polynomial-time[14] machine P^*, for all sufficiently long $x \notin S$, and for all $w \in \{0,1\}^*$, the verifier V rejects with probability at least $\frac{1}{2}$, after interacting with $P^*(w)$ on common input x.

Both conditions can be rephrased by using (non-uniform) families of circuits of polynomial size. Argument systems are adequate for modeling the behavior of parties in a real-life setting. Under strong intractability assumptions, argument systems exhibit advantages over interactive proof systems.[15] Let us start by stating these assumptions.

Definition 2.18 (Collision-Free Hashing): *Consider a family of hash functions, indexed by strings, $F \stackrel{\text{def}}{=} \{f_\alpha : \{0,1\}^{2|\alpha|} \mapsto \{0,1\}^{|\alpha|}\}_\alpha$, so that there exists a polynomial-time algorithm for evaluating F (i.e., on input α and x returns $f_\alpha(x)$). The family F is called* collision-free *w.r.t. complexity $c(\cdot)$ if for every non-uniform family of circuits $\{C_n\}$ with size bounded by $c(\cdot)$, and all sufficiently large n's, the probability that C_n, given a uniformly chosen $\alpha \in \{0,1\}^n$, outputs a pair (x,y) so that $f_\alpha(x) = f_\alpha(y)$, is bounded above by $1/c(n)$. The family F is called* collision-free *if it is collision-free w.r.t. all polynomials, and is called* strongly collision-free *if, for some $\epsilon > 0$, it is collision-free w.r.t. the function $f(n) \stackrel{\text{def}}{=} 2^{n^\epsilon}$.*

---

[14] Again, this means a running time polynomial in the length of the common input.

[15] Below, we consider the expressing power of both models. An additional advantage of argument systems is that, under strong intractability assumptions, there exist *perfect* zero-knowledge arguments (rather than *computational* zero-knowledge interactive proofs) for any set in $\mathcal{NP}$ [77]. Recall that perfect zero-knowledge proofs may exist only for sets in $\mathcal{IP}(2) \cap \text{co}\mathcal{IP}(2)$ [153, 2].

Collision-free functions exist assuming the intractability of factoring integers (i.e., in polynomial time). Strong collision-free functions exist if n-bit long integers cannot be factored in time 2^{n^ϵ}, for some $\epsilon > 0$.

Theorem 2.19 [232]: *Let $L \in \mathcal{NP}$ and assume the existence of collision-free functions* (resp., strong collision-free functions). *Then, for every $\epsilon > 0$, there exists an argument system for L in which the randomness and communication complexities, on inputs of length n, are both bounded by n^ϵ* (resp., by poly($\log(n)$)). *Furthermore, the computational complexity of the verifier is quadratic in the length of the input.*

The theorem is proved by combining techniques from Cryptography with constructions of PCP systems (as of Theorem 2.12). Specifically, the prover commits to a proof-oracle of the PCP system using an "authentication tree" in which each node holds the hash value (under a collision-free function) of its two children. To inspect a specific leaf, it suffices to obtain the values of all nodes along the path from the root to this leaf as well as the values of their immediate children.

We stress that Theorem 2.19 is meaningful also in case $L \in \mathcal{P}$; in particular, it offers quadratic verification time, independently of the (possibly higher) deterministic complexity of the set. Interestingly, the results of Theorem 2.19 are unlikely for interactive proof systems, due to the following:

Proposition 2.20 [176]: *Suppose that L has an interactive proof system in which both the randomness and communication complexities are bounded by an integer function $c(\cdot)$. Then $L \in$ Dtime($2^{O(c(\cdot)+\log(\cdot))}$).*

Further results of similar nature are also presented in [176].

Proof Idea: Consider the tree of all possible executions (see Definition C.2). □

CS-Proof Systems. The definition of a CS-proof system is derived from the definition of an interactive proof system analogously to the way the definition of an argument system is derived. The difference is that here the potential provers are *uniform* probabilistic machines, with no auxiliary inputs, running in time polynomial *in the deterministic complexity of the set*. A result analogous to Theorem 2.19 is obtainable also in the current setting. Specifically,

Theorem 2.21 [267]: *Let $S \in \mathcal{EXP}$. Then, assuming the existence of strong collision-free functions, there exists a CS-proof system for S. Furthermore, fixing any decision procedure D for the set S, the following holds, for each $x \in S$,*

1. *The running-time of the verifier, on common input x, is quadratic in the length of the input and poly-logarithmic in the running time of D on x.*
2. *The running-time of the (prescribed) prover, on common input x, is polynomial in the running time of D on x.*

In fact, the above additional requirements are incorporated into the actual definition of CS-proofs in [267]. Thus, the actual definition of CS-proofs yields a notion of proof systems in which proving is not much harder than deciding, in a strong "pointwise" sense.

2.5.2 Non-Interactive Proofs

The class $\mathcal{IP}(1)$ may be considered *the real* model of non-interactive probabilistic proof systems. It extends $\mathcal{NP}$ in allowing the verifier to toss coins while examining a candidate proof of polynomial (in the assertion) length. Two more interesting models are discussed below.

Non-Interactive Zero-Knowledge Proofs (NIZK). Actually the term "non-interactive" is somewhat misleading. The model, introduced in [66], consists of three entities: a prover, a verifier and a uniformly selected sequence of bits (which can be thought of as being selected by a trusted third party). Both verifier and prover can read the random sequence, and each can toss additional coins. The interaction consists of a single message sent from the prover to the verifier, who then is left with the decision (whether to accept or not). Based on some reasonable complexity assumptions, one may construct non-interactive zero-knowledge proof systems for every NP-set (cf., [66, 143, 233]).

Non-Interactive CS-proofs. Actually, [267] presents two different models of non-interactive CS-proofs.

1. Ordinary non-interactive CS-proofs relate to (interactive) CS-proofs (as presented above) analogously to the relation of ordinary non-interactive proofs to interactive proofs. That is, both the prover and the verifier have access to a random bit sequence (of polynomial length). A plausibility argument towards the existence of such non-trivial CS-proofs is given in [267]; but it is an important open problem (e.g., see applications in [267]), whether such CS-proofs can be constructed, say for $\mathcal{NP}$, based on standard intractability assumptions.
2. CS-proofs in the Random Oracle Model (i.e., both parties have access to a random oracle). The existence of such CS-proofs for any set in $\mathcal{EXP}$ is proven in [267] (without relying on any complexity assumptions).

2.5.3 Proofs of Knowledge

The concept of a proof of knowledge, introduced in [199], is very appealing; yet, its precise formulation is much more complex than one may expect (cf. [36]). Loosely speaking, a knowledge-verifier for a relation R guarantees the existence of a "knowledge extractor" that on input x and access to any interactive machine P^* outputs a y, so that $(x, y) \in R$, within complexity (inversely) related to the probability that the verifier accepts x when interacting with P^*. By convincing such a knowledge-verifier, on common input x, one proves that he knows a y so that $(x, y) \in R$. Clearly, any NP-verifier (i.e., accepting x iff it receives an NP-witness w.r.t R) is a knowledge-verifier for the corresponding NP-relation. More interestingly, the (zero-knowledge) protocol which results by successively applying Construction 2.7 sufficiently many time constitutes a "proof of knowledge" of a 3-coloring of the input graph.

2.5.4 Refereed Games

The following notion of *refereed games* seems related to Multi-Prover Interactive Proof (MIP) systems, but is actually very different both conceptually and technically. Whereas in MIP the two proves try to coordinate their strategies so as to convince the verifier of the validity of a given assertion, in a Refereed Game one party tries to convince the verifier (called the *referee*) that the assertion is valid whereas the other tries to refute the assertion [146, 142]. Thus, the correctness of the referee's decision depends on the assumption that the party which is right plays well (if not optimally). As in all proof systems discussed above, also here the referee (or verifier) employs a probabilistic polynomial-time strategy. The refereed game may either be a game of full information (i.e., each of the competing players sees all messages sent) or be a game of partial information (i.e., each obtains only the messages sent to it by the referee). It turns out that the latter are more powerful [142].

2.6 Concluding Remarks

In this section we compare the various proof systems presented above, provide a historical account of their evolution, and propose some open problems.

2.6.1 Comparison Among the Various Notions

All the above variants of probabilistic proof systems are aimed at capturing central aspects of the intuitive notion of efficient proof systems. Although the alternative formulations are sometimes introduced using the same generic phrases, they are actually very different in motivation, applications and expressive power. The objective of this section is to try to clarify these differences.

In Figure 2.1, we have tried to summarize the differences between the various notions of efficient proof systems. The class $\mathcal{NP}$ has been omitted for obvious reasons. We view $\mathcal{IP}$ as *the* natural *generalization* of $\mathcal{NP}$, obtained by relaxing the notion of efficient computation so that probabilism and interaction are allowed. Except for the negligible probability of error, which can

	IP	arguments	CS-proof	PCP	MIP
restrictions on prover	none	poly-time + aux. input	polynomial in Dec. time	memoryless (i.e., oracle)	split entity
motivation (as we see it)	generalize NP	restrict IP (see Remark 1)		augment NP	see Remark 2
expressive power	$\mathcal{PSPACE}$	$\mathcal{IP}(1) \subseteq \mathcal{PH}$	$\mathcal{EXP}^{17}$	scalable: Ntime($2^{l(n)}$), for rnd+query $= O(l(n))$	

Fig. 2.1. Comparison of various proof systems

be controlled by the verifier, the original flavor of a proof is maintained. Also, we view $\mathcal{PCP}(\log, O(1))$ as an *augmentation* of $\mathcal{NP}$ with the extra property of allowing a hasty verifier to take its chances and verify the proof in a super-fast manner.[16] In contrast, the two notions of computationally sound proof systems (i.e., arguments and CS-proofs) deviate significantly from the conservative approach of absolute proofs. Yet, computational soundness seems adequate in most practical settings. The only word of warning is that typical results in these latter settings depend on intractability assumptions, and when evaluating such results one should not ignore the relative severeness of these assumptions.

Remark 1: Arguments and CS-proof systems are derived by imposing computational restrictions on the potential provers in both the completeness and soundness conditions. In both cases the motivation for these restrictions is to obtain properties that interactive proofs do not (seem to) have. In the case of argument systems the advantageous properties are very low communication complexity and *perfect* zero-knowledge (for $\mathcal{NP}$). Interestingly, the expressive power of the system does not increase in this case (but rather decreases). In the case of CS-proof systems the advantageous property is the linking of the complexity of proving to the complexity of deciding. Interestingly, the expressive power of the system seems to increase as well (unless $\mathcal{PSPACE} = \mathcal{EXP}$).

Remark 2: The MIP model indeed generalizes the IP model. However, in our opinion, this generalization is less natural than the generalization of NP to IP. As far as we are concerned, the MIP model is justified by cryptographic applications (see subsection on MIP). (The transformations between MIP systems and PCP systems does not mean that the motivation of one model can be moved to the other.)

---

[16] Recall that the oracle guaranteed by the completeness condition (of the definition of $\mathcal{PCP}(\log, O(1))$) provides a standard NP-proof. The additional feature of allowing hasty probabilistic verification accounts for the term 'augmentation' used above.

[17] Depending on (strong) intractability assumptions.

Remark 3: As mentioned above, the error probability can be decreased in all these probabilistic proof systems, by using *sequential* repetitions. Error reduction by *parallel* repetitions is more problematic, with the exception of (plain) interactive proof systems (alas even in this vanilla case, parallel repetition is less trivial to analyze than sequential repetition; see Appendix C.1). A Parallel Repetition Theorem for one-round multi-party interactive proofs was proven by Raz [309] (cf., Appendix C.1 and [136] for further discussion). Zero-knowledge is not preserved, in general, under parallel repetition (cf., [179]). A recent study shows that parallel repetition is problematic also in case of computationally-sound proof systems (cf., [46]).

2.6.2 The Story

In this section we provide a historical account of the evolution of probabilistic proof systems. We focus on the main conceptual discoveries, neglecting many of the technical contributions which played an important role in the development of the area.

The introduction of interactive proofs and zero-knowledge proofs. The story begins with Goldwasser, Micali and Rackoff who sought a general setting for their novel notion of zero-knowledge [199]. The choice fell on proof systems – as capturing a fundamental activity which takes place in a cryptographic protocol. Motivated by the desire to formulate the most general type of "proofs" that may be used within cryptographic protocols, Goldwasser, Micali and Rackoff introduced the notion of an *interactive proof system* [199]. Although the main thrust of their paper is the introduction of a special type of interactive proofs (i.e., ones that are *zero-knowledge*), the possibility that interactive proof systems may be more powerful from NP-proof system has been pointed out in [199].

Independently of [199],[18] Babai suggested a different formulation of interactive proofs, which he called *Arthur-Merlin Games* [23]. Syntactically, Arthur-Merlin Games are a restricted form of interactive proof systems, yet it was subsequently shown that these restricted systems are as powerful as the general ones (cf., [203]). Babai's motivation was to place a group-theoretic problem, previously placed in $\mathcal{NP}$ under some group-theoretic assumptions, "as close to $\mathcal{NP}$ as possible" without using any assumptions. Interestingly, Babai underestimated the expressive power of interactive proof systems, conjecturing that the class of sets possessing such proof systems (even with an unbounded number of message-exchange rounds) is "very close" to $\mathcal{NP}$.

Discovering the power of zero-knowledge proofs. The first evidence of the surprising power of interactive proofs was given by Goldreich, Micali, and Wigderson, who presented an interactive proof system for Graph

---

[18] Although both [199] and [23] have appeared in the same conference (i.e., *17th STOC*, 1985), early versions of [199] have existed as early as 1982, and were rejected three times from major conferences (i.e., *FOCS83*, *STOC84*, and *FOCS84*).

Non-Isomorphism [185], a set not known to be in $\mathcal{NP}$. More importantly, this paper has demonstrated the generality and wide applicability of zero-knowledge proofs. Assuming the existence of one-way function, it was shown how to construct zero-knowledge interactive proofs for any set in $\mathcal{NP}$. This result has had a dramatic impact on the design of cryptographic protocols (cf., [186]). In addition, this result has called attention to the then new notion of interactive proof systems (since zero-knowledge NP-proofs could exist only in a trivial sense [187]).

Multi-Prover Interactive Proof Systems. A generalization of interactive proofs to *multi-prover interactive proofs* has been suggested by Ben-Or, Goldwasser, Kilian and Wigderson [58]. Again, the main motivation came from zero-knowledge aspects; specifically, introducing multi-prover zero-knowledge proofs for $\mathcal{NP}$ without relying on intractability assumptions. Yet, the complexity theoretic prospects of the new class, denoted $\mathcal{MIP}$, have not been ignored. A more appealing, to our taste, formulation of the class $\mathcal{MIP}$ has been presented in [154]. The latter formulation exactly coincides with the formulation now known as *probabilistically checkable proofs* (i.e., $\mathcal{PCP}$).

Algebraic Methods Demonstrate the Power of Interactive Proofs. The amazing power of interactive proof systems has been demonstrated by using algebraic methods. The basic technique has been introduced by Lund, Fortnow, Karloff and Nisan, who applied it to show that the polynomial-time hierarchy (and actually $\mathcal{P}^{\#\mathcal{P}}$) is in $\mathcal{IP}$ [255]. Subsequently, Shamir used the technique to show that $\mathcal{IP} = \mathcal{PSPACE}$ [325], and Babai, Fortnow and Lund used it to show that $\mathcal{MIP} = \mathcal{NEXP}$ [24].

The technique of Lund *et. al.* [255] has been inspired by ideas coming from works on "program checking" (cf., [70]). In particular, their interactive proof system for the permanent combines the "self-correcting" procedure for the permanent (which represents the permanent as a multi-linear polynomial) of [248], and the "downwards self-reducibility" procedure of [69]. Another idea that is implicit in [255] and made explicit in the subsequent works of [325, 24] is the representation, introduced in [31], of Boolean formulae as multi-linear polynomials.

It may be of interest to note that the technique of Lund *et. al.* has been first applied in the context of multi-prover interactive proofs, yielding $\mathcal{P}^{\#\mathcal{P}} \subseteq \mathcal{MIP}$, and that the result quoted above (concerning $\mathcal{IP}$) followed later. Hence, $\mathcal{MIP}$ has played a role in the historical development leading to the characterization of $\mathcal{IP}$.

Scaling Down the BFL Proof System Yields a New Class. The abovementioned multi-prover proof system of Babai, Fortnow and Lund [24] (hereafter referred to as the BFL proof system) has been the starting point for fundamental developments regarding $\mathcal{NP}$. The first development was the discovery that the BFL proof system can be "scaled-down"[19] from $\mathcal{NEXP}$

---

[19] The term "scaled-down" is used here as a (standard) technical term. Doing so, I do *not* mean to underestimate the technical difficulty of obtaining these results.

to $\mathcal{NP}$. This important discovery was made independently by two sets of authors: Babai, Fortnow, Levin and Szegedy [25] and Feige, Goldwasser, Lovasz and Safra [138].[20] However, the manner in which the BFL proof is scaled-down is different in the two papers, and so are the consequences of the scaling-down.

Babai *et. al.* [25] start by considering only inputs encoded using a special error-correcting code. The encoding of strings, relative to this error-correcting code, can be computed in polynomial time. They presented a polynomial-time algorithm that transforms NP-witnesses (to inputs in a set $S \in \mathcal{NP}$) into *transparent proofs* that can be verified as vouching for the correctness of the encoded assertion in (probabilistic) poly-logarithmic time (by a Random Access Machine). (The fact that the verification procedure never reads the entire "proof" should not come as a surprise, as the procedures of [255, 325, 24] also have this property.) Thus, once "statements" and "proofs" are in the right (error-correcting) form, verification is "super-fast." Babai *et. al.* [25] stress the practical aspects of transparent proofs – specifically, for rapidly checking transcripts of long computations.

In the proof system of Babai *et. al.* [25] the total running time of the verifier is reduced (i.e., "scaled-down") to poly-logarithmic. In contrast, in the proof system of Feige *et. al.* [138, 139] the verifier stays polynomial-time and only two more refined complexity measures, specifically the randomness and query complexities, are reduced to poly-logarithmic. This eliminates the need to assume that the input is in a special error-correcting form, and yields a more appealing (i.e., less cumbersome) complexity class. This complexity class is a refinement of the class introduced in [154]. The refinement is obtained by specifying the randomness and query complexities. Namely, $\mathcal{PCP}(r(\cdot), q(\cdot))$ denotes the class of sets having probabilistically checkable proofs in which, on input x, the verifier tosses at most $r(|x|)$ coins and makes at most $q(|x|)$ (Boolean) queries to the proof. Hence, whereas the result of Babai *et. al.* [24] can be restated as

$$\mathcal{NEXP} = \mathcal{PCP}(\text{poly}, \text{poly}), \tag{2.2}$$

the result of Feige *et. al.* [139] is restated as

$$\mathcal{NP} \subseteq \mathcal{PCP}(f(\cdot), f(\cdot)), \qquad \text{where } f(n) = O(\log n \cdot \log \log n). \tag{2.3}$$

It should be stressed that the result of Babai *et. al.* [25] also implies

$$\mathcal{NP} \subseteq \mathcal{PCP}(\texttt{log}, \texttt{polylog}). \tag{2.4}$$

Interest in the new complexity class became immense since Feige *et. al.* [138, 139] demonstrated its relevance to proving the intractability of approximating some combinatorial problems (specifically, MaxClique). When using the

---

[20] At a later stage, Szegedy improved the randomness and query complexities of the system in [138] and joined the latter paper, which has appeared as [139].

PCP–MaxClique connection established by Feige *et. al.*, the randomness and query complexities of the verifier (in a pcp system for an NP-complete set) relate to the strength of the negative results obtained for approximation problems. This fact provided a very strong motivation for trying to reduce these complexities and obtain a tight characterization of $\mathcal{NP}$ in terms of $\mathcal{PCP}(\cdot, \cdot)$.

Tightening the Relation between NP and PCP. Once the work of Feige *et. al.* [139] had been presented, the challenge was clear – showing that $\mathcal{NP}$ equals $\mathcal{PCP}(\log, \log)$. This challenge was met by Arora and Safra [20]. The proof system they constructed is very complex, involving recursive use of proof systems and concatenation tests that are much more efficient than the length of strings being tested. (Interestingly, the idea of encoding inputs in an error-correcting form, as suggested in [25], is essential to make this recursion work.) Actually, Arora and Safra showed that

$$\mathcal{NP} = \mathcal{PCP}(\log, f(\cdot)), \qquad \text{where } f(n) = o(\log n). \tag{2.5}$$

Hence, a new challenge arose, namely, further reducing the query complexity – in particular to a constant – while maintaining the logarithmic randomness complexity. Again, additional motivation for this challenge came from the relevance of such a result to the study of approximation problems. The new challenge was met by Arora, Lund, Motwani, Sudan and Szegedy [19], and is captured by the equation

$$\mathcal{NP} = \mathcal{PCP}(\log, O(1)). \tag{2.6}$$

In addition to building on the ideas of Arora and Safra [20], the above result of [19] utilizes ideas and techniques from the works on self-testing/self-correcting [69], degree-tests for multi-variant polynomials [161, 314], and parallelization of multi-prover proof systems [239].

Derandomization techniques were extensively used in the above as well as subsequent works. In particular, pairwise-independent sampling [103] is essential to [25] (and instrumental for obtaining the best bounds in [139]), small-bias spaces [274] are implicit in [25, 139], and random walks on expander graphs [4] are used from [20] onwards,

Computationally-Sound Proof Systems. Argument systems were defined in 1986 by Brassard, Chaum and Crépeau [77], but their complexity-theoretic significance became apparent only in 1992. This happened when Kilian, using early results on $\mathcal{PCP}$ (due to [25, 139]), showed that, under some reasonable intractability assumptions, every set in $\mathcal{NP}$ has a computationally-sound proof in which the randomness and communication complexities are poly-logarithmic [232]. Consequently, Micali suggested three new types of computationally-sound proof systems which he called CS-proofs [266, 267].

Other Types of Proof Systems. The setting of non-interactive proofs was first introduced by Blum, Feldman and Micali [66]. The concept of proofs of knowledge was introduced in the paper of Goldwasser, Micali and Rackoff [199], and given a satisfactory formal treatment in [36].

2.6.3 Open Problems

We disagree with the general sentiment according to which the nature of the various probabilistic proof systems is well understood by now. In contrast, we point out several important directions for future research:

1. *The structure of the IP Hierarchy:* The relatively early discovery of the exact expressive power of interactive proofs (i.e., Theorem 2.4) caused researchers to forget that except for the Linear Speed-up Theorem of [27] we know little about the impact of the number of interactions on the expressive power.

2. *A non-tricky proof of* $\mathcal{IP} = \mathcal{PSAPCE}$: It seems strange that the proof of such a fundamental result about computation has to rely on mysterious algebraic tricks. Things become even worse when one gets to the proof of the PCP Characterization of $\mathcal{NP}$ (i.e., Theorem 2.12). We refer to the key role of polynomials in the above constructions. We consider it important to obtain an alternative proof of co$\mathcal{NP} \subseteq \mathcal{IP}$; a proof in which all the underlying ideas can be presented at an abstract level.

3. *The power of the prover in interactive proofs:* We ask how powerful should be a prover which is able to convince a verifier for a set S. The question is aimed at characterizing classes of sets for which relative efficient provers exists, where we refer to either the second or the third notion of relative efficiency mentioned in Section 2.2.5. (For the first notion the answer is trivial.)

4. *Simplifying the proof of the PCP Characterization of NP:* It is very annoying that the current proof is so complex. One question is whether the proof composition paradigm is indeed essential. However, given the role it plays in subsequent developments, we are tempted to let it stay. In such a case one is left with the question of how to construct a $\mathcal{PCP}(\log, \text{poly}(\log))$ system for $\mathcal{NP}$, having the extra properties required in the proof composition (see proof sketch above). Specifically, we refer to the requirement that the verifier makes a constant number of queries to a multi-valued oracle. Thus, given an arbitrary $\mathcal{PCP}(\log, \text{poly}(\log))$ system for $\mathcal{NP}$, one wishes to construct a system in which the latter property holds. We seek an alternative way of obtaining such a "parallelization" – one which does not rely on non-abstractable algebraic creatures (like polynomials). A first step towards this partial goal was taken in [191]: It was shown how to construct an efficient low-degree test which utilizes a specific simple/inefficient low-degree test as a subroutine, and reduces it error probability via a parallelization which is analyzed using a new "combinatorial consistency lemma".

5. *The power of ordinary non-interactive CS-proofs*: Positive results regarding CS-proofs are known only in the interactive model and in the Random Oracle Model [266, 267]. Any non-trivial positive results, under standard intractability assumptions, for the ordinary non-interactive model will be of interest.

6. *Computational ZK proofs vs Perfect ZK arguments*: Computational zero-knowledge proofs and perfect zero-knowledge arguments seem to be dual terms. However, the former can be constructed for $\mathcal{NP}$ based on any one-way function [185], whereas the latter can be constructed (for $\mathcal{NP}$) based on one-way permutations [77, 275]. Is this discrepancy fundamental?

7. *Constant-round zero-knowledge proofs for NP*: The known *constant-round* zero-knowledge proofs for $\mathcal{NP}$ use *expected polynomial-time* simulators, rather than *strict* polynomial-time ones (cf., [178]). Can this annoying technicality be removed?

Acknowledgments

I am grateful to Shafi Goldwasser for suggesting the essential role of randomness as the unifying theme for this exposition. Thanks also to Leonid Levin, Dana Ron, Madhu Sudan, Luca Trevisan and Uri Zwick for commenting on earlier versions of this chapter.

3. Pseudorandom Generators

If two objects are indistinguishable, in what sense are they different?

The author, failing to recall a suitable quote (1997).

Summary – A fresh view at the *question of randomness* was taken in the theory of computing: It has been postulated that a distribution is pseudorandom if it cannot be told apart from the uniform distribution by an efficient procedure. The paradigm, originally associating efficient procedures with polynomial-time algorithms, has been applied also with respect to a variety of limited classes of such distinguishing procedures. Starting with the general paradigm, we survey the archetypical case of pseudorandom generators (withstanding any polynomial-time distinguisher), as well as generators withstanding space-bounded distinguishers, the derandomization of complexity classes such as $\mathcal{BPP}$, and some special-purpose generators.

3.1 Introduction

The second half of this century has witnessed the development of three theories of randomness, a notion which has been puzzling thinkers for ages. The first theory (cf., [109]), initiated by Shannon [322], is rooted in probability theory and is focused at distributions which are not perfectly random. Shannon's Information Theory characterizes perfect randomness as the extreme case in which the *information contents* is maximized (and there is no redundancy at all). Thus, perfect randomness is associated with a unique distribution – the uniform one. In particular, by definition, one cannot generate such perfect random strings from shorter random seeds.

The second theory (cf., [243, 246]), due to Solomonov [333], Kolmogorov [235] and Chaitin [93], is rooted in computability theory and specifically in the notion of a universal language (equiv., universal machine or computing device). It measures the complexity of objects in terms of the shortest program (for a fixed universal machine) which generates the object. Like

Shannon's theory, Kolmogorov Complexity is quantitative and perfect random objects appear as an extreme case. However, in this approach one may say that a single object, rather than a distribution over objects, is perfectly random. Still, Kolmogorov's approach is inherently intractable (i.e., Kolmogorov Complexity is uncomputable), and – by definition – one cannot generate strings of high Kolmogorov Complexity from short random seeds.

The third theory, initiated by Blum, Goldwasser, Micali and Yao [198, 71, 351], is rooted in complexity theory and is the focus of this chapter. This approach is explicitly aimed at providing a notion of perfect randomness which nevertheless allows to efficiently generate perfect random strings from shorter random seeds. The heart of this approach is the suggestion to view objects as equal if they cannot be told apart by any efficient procedure. Consequently a distribution which cannot be efficiently distinguished from the uniform distribution will be considered as being random (or rather called pseudorandom). Thus, randomness is not an "inherent" property of objects (or distributions) but rather relative to an observer (and its computational abilities). To demonstrate this approach, let us consider the following mental experiment.

> Alice and Bob play "head or tail" in one of the following four ways. In all of them Alice flips an unbiased coin and Bob is asked to guess its outcome *before* the coin hits the floor. The alternative ways differ by the knowledge Bob has before making his guess. In the first alternative, Bob has to announce his guess before Alice flips the coin. Clearly, in this case Bob wins with probability $1/2$. In the second alternative, Bob has to announce his guess while the coin is spinning in the air. Although the outcome is *determined in principle* by the motion of the coin, Bob does not have accurate information on the motion and thus we believe that also in this case Bob wins with probability $1/2$. The third alternative is similar to the second, except that Bob has at his disposal sophisticated equipment capable of providing accurate *information* on the coin's motion as well as on the environment effecting the outcome. However, Bob cannot process this information in time to improve his guess. In the fourth alternative, Bob's recording equipment is directly connected to a *powerful computer* programmed to solve the motion equations and output a prediction. It is conceivable that in such a case Bob can improve substantially his guess of the outcome of the coin.

We conclude that the randomness of an event is relative to the information and computing resources at our disposal. Thus, a natural concept of pseudorandomness arises – a distribution is *pseudorandom* if no efficient procedure can distinguish it from the uniform distribution, where efficient procedures are associated with (probabilistic) polynomial-time algorithms. This notion of pseudorandomness is indeed the most fundamental one, and much of this chapter is focused on it. Weaker notions of pseudorandomness arise as well –

they refer to indistinguishability by weaker procedures such as space-bounded algorithms, constant-depth circuits, etc. Stretching this approach even further one may consider algorithm which are designed on purpose so not to distinguish even weaker forms of "pseudorandom" sequences from random ones (such algorithms arise naturally when trying to convert some natural randomized algorithm into deterministic ones; see Section 3.6).

The above discussion has focused on one aspect of the pseudorandomness question – the resources or type of the observer (or potential distinguisher). Another important question is whether such pseudorandom sequences can be generated from much shorter ones, and at what cost (or complexity). A natural answer is that the generation process has to be at least as efficient as the efficiency limitations of the distinguisher. Coupled with the above-mentioned strong notion of pseudorandomness, this yields the archetypical notion of pseudorandom generators – these operating in polynomial-time and producing sequences which are indistinguishable from uniform ones by *any* polynomial-time observer. Such pseudorandom generators allow to reduced the randomness complexity of *any efficient application*, and are thus of great relevance to randomized algorithms, cryptography and complexity theory (see Section 3.3). Interestingly, there are important reasons to consider also an alternative which seems less natural; that is, allow the generator to use more resources (e.g., time or space) than the observer it tries to fool. Indeed, this makes the task of designing pseudorandom generators easier, but the usefulness of such generators has to be demonstrated – as done in Sections 3.4 through 3.6.

Organization. In Section 3.2 we present the general paradigm underlying all the various notions of pseudorandom generators. The archetypical case (of generators operating in polynomial-time and fooling all polynomial-time distinguishers) is discussed in Section 3.3. We then turn to the alternative notions of pseudorandom generators: Generators which work in time exponential in the length of the seed, and suffice for the derandomization of complexity classes such as $\mathcal{BPP}$, are discussed in Section 3.4; Pseudorandom generators in the domain of space-bounded computations are discussed in Section 3.5; and special-purpose generators are discussed in Section 3.6. Concluding remarks appear in Section 3.7.

For an alternative presentation, which focuses on the archetypical case and provides more details on it, the reader is referred to [170, Chap. 3].

3.2 The General Paradigm

A generic formulation of pseudorandom generators consists of specifying three fundamental aspects – the *stretching measure* of the generators; the class of distinguishers that the generators are supposed to fool (i.e., the algorithms with respect to which the *computational indistinguishability* requirement should hold); and the resources that the generators are allowed to use (i.e., their own *computational complexity*).

Stretching function: A necessary requirement from any notion of a pseudorandom generator is that it is a deterministic algorithm which stretches short strings, called *seeds*, into longer output sequences. Specifically, it stretches k-bit long seeds into $\ell(k)$-bit long outputs, where $\ell(k) > k$. The function ℓ is called the *stretching measure* (or *stretching function*). In some settings the specific stretching measure is immaterial (e.g., see Section 3.3).

Computational Indistinguishability: A necessary requirement from any notion of a pseudorandom generator is that it "fools" some non-trivial algorithms. That is, any algorithm taken from some class of interest cannot distinguish the output produced by the generator (when the generator is fed with a uniformly chosen seed) from a uniformly chosen sequence. Typically, we consider a class $\mathcal{D}$ of distinguishers and a class $\mathcal{F}$ of noticeable functions, and require that the generator G satisfies the following: For any $D \in \mathcal{D}$, any $f \in \mathcal{F}$, and for all sufficiently large k's

$$| \Pr[D(G(U_k)) = 1] - \Pr[D(U_{\ell(k)}) = 1] | \; < \; f(k)$$

where U_n denotes the uniform distribution over $\{0,1\}^n$ and the probability is taken over U_k (resp., $U_{\ell(k)}$) as well as over the coin tosses of algorithm D in case it is probabilistic.[1] The archetypical choice is that $\mathcal{D}$ is the set of probabilistic polynomial-time algorithms, and $\mathcal{F}$ is the set of functions which are the reciprocal of some positive polynomial.

Complexity of Generation: The archetypical choice is that the generator has to work in polynomial-time (in length of its input – the seed). Other choices will be discussed as well. We note that placing no computational requirements on the generator (or, alternatively, putting very mild requirements such as a double-exponential running-time upper bound), yields "generators" which can fool any subexponential-size circuit family [180].

Notational conventions. We will consistently use k to denote the length of the seed of a pseudorandom generator, and $\ell(k)$ to denote the length of the corresponding output. In some cases, this makes our presentation a little more cumbersome (as a natural presentation may specify some other parameters and let the seed-length be a function of these). However, our choice has the advantage of focusing attention on the fundamental parameter of pseudorandom generation – the length of the random seed. Whenever a pseudorandom generator is used to "derandomize" an algorithm, n will denote the length of the input to this algorithm, and k will be selected as a function of n.

---

[1] Thus, we require certain functions (i.e., the absolute difference between the above probabilities), to be smaller than any noticeable function *on all but finitely many integers*. We call such functions negligible. Note that a function may be neither noticeable nor negligible (e.g., it may be smaller than any noticeable function on infinitely many values and yet larger than some noticeable function on infinitely many other values).

3.3 The Archetypical Case

As stated above, the most natural notion of a pseudorandom generator refers to the case where both the generator and the potential distinguisher work in polynomial-time. Actually, the distinguisher is more complex than the generator: The generator is a fixed algorithm working within *some fixed* polynomial-time, whereas a potential distinguisher is *any* algorithm which runs in polynomial-time. Thus, for example, the distinguisher *may* always run in time cubic in the running-time of the generator. Furthermore, to facilitate the development of this theory, we allow the distinguisher to be probabilistic (whereas the generator remains deterministic as above). In the role of the set of noticeable functions we consider all functions which are the reciprocal of some positive polynomial.[2] This choice is naturally coupled with the association of efficient computation with polynomial-time algorithms: An event which occurs with noticeable probability occurs almost always when the experiment is repeated a "feasible" (i.e., polynomial) number of times. This discussion leads to the following instantiation of the generic framework presented above –

Definition 3.1 (pseudorandom generator – archetypical case [71, 351]): *A deterministic polynomial-time algorithm G is called a* pseudorandom genera-tor *if there exists a* stretching function, $\ell: \mathbb{N} \mapsto \mathbb{N}$, *so that for any probabilistic polynomial-time algorithm D, for any positive polynomial p, and for all sufficiently large k's*

$$| \Pr[D(G(U_k)) = 1] - \Pr[D(U_{\ell(k)}) = 1] | < \frac{1}{p(k)}$$

where U_n denotes the uniform distribution over $\{0,1\}^n$ and the probability is taken over U_k (resp., $U_{\ell(k)}$) as well as over the coin tosses of D.

Thus, pseudorandom generators are efficient (i.e., polynomial-time) deterministic programs which expand short randomly selected seeds into longer pseudorandom bit sequences, where the latter are defined as computationally indistinguishable from truly random sequences by efficient (i.e., polynomial-time) algorithms. It follows that any efficient randomized algorithm maintains its performance when its internal coin tosses are substituted by a sequence generated by a pseudorandom generator. That is,

Construction 3.2 (typical application of pseudorandom generators): *Let A be a probabilistic algorithm, and $\rho(n)$ denote a (polynomial) upper bound on*

---

[2] The definition below asserts that the distinguishing gap of certain machines must be smaller than the reciprocal of any positive polynomial for all but finitely many n's. Such functions are called *negligible*. See Footnote 1. The notion of negligible probability is robust in the sense that an event which occurs with negligible probability occurs with negligible probability also when the experiment is repeated a "feasible" (i.e., polynomial) number of times.

its randomness complexity. Let $A(x, r)$ denote the output of A on input x and coin tosses sequence $r \in \{0, 1\}^{\rho(|x|)}$. Let G be a pseudorandom generator with stretching function $\ell : \mathbb{N} \mapsto \mathbb{N}$. Then A_G is a randomized algorithm which on input x, proceeds as follows. It sets $k = k(|x|)$ to be the smallest integer such that $\ell(k) \geq \rho(|x|)$, uniformly selects $s \in \{0, 1\}^k$, and outputs $A(x, r)$, where r is the $\rho(|x|)$-bit long prefix of $G(s)$.

We show that it is infeasible to find long x's on which the *noticeable behavior* of A_G is different from the one of A, although A_G may use much fewer coin tosses. That is

Proposition 3.3 *Let A and G be as above. Then for every pair of probabilistic polynomial-time algorithms, a finder F and a distinguisher D, every positive polynomial p and all sufficiently long n's*

$$\sum_{x \in \{0,1\}^n} \Pr[F(1^n) = x] \cdot \Delta_{A,D}(x) \; < \; \frac{1}{p(n)}$$

where $\Delta_{A,D}(x) \stackrel{\text{def}}{=} |\Pr[D(x, A(x, U_{\rho(n)})) = 1] - \Pr[D(x, A_G(x, U_{k(n)})) = 1]|$ and the probabilities are taken over the U_m's as well as over the coin tosses of F and D.

The proposition is proven by showing that a triplet (A, F, D) violating the claim can be converted into an algorithm D' which distinguishes the output of G from the uniform distribution, in contradiction to the hypothesis. Analogous arguments are applied whenever one wishes to prove that an efficient randomized process (be it an algorithm as above or a multi-party computation) preserves its behavior when one replaces true randomness by pseudorandomness as defined above. Thus, given pseudorandom generators with large stretching function, *one can considerably reduce the randomness complexity in any efficient application.*

3.3.1 A Short Discussion

Randomness is playing an increasingly important role in computation: It is frequently used in the design of sequential, parallel and distributed algorithms, and is of course central to cryptography. Whereas it is convenient to design such algorithms making free use of randomness, it is also desirable to minimize the usage of randomness in real implementations. Thus, pseudorandom generators (as defined above) are a key ingredient in an "algorithmic tool-box" – they provide an automatic compiler of programs written with free usage of randomness into programs which make an economical use of randomness.

Indeed, "pseudo-random number generators" have appeared with the first computers. However, typical implementations use generators which are not pseudorandom according to the above definition. Instead, at best, these generators are shown to pass SOME ad-hoc statistical test (cf., [234]). However, the fact that a "pseudo-random number generator" passes some statistical tests, does not mean that it will pass a new test and that it is good for a future (untested) application. Furthermore, the approach of subjecting the generator to some ad-hoc tests fails to provide general results of the type stated above (i.e., of the form "for ALL practical purposes using the output of the generator is as good as using truly unbiased coin tosses"). In contrast, the approach encompassed in Definition 3.1 aims at such generality, and in fact is tailored to obtain it: The notion of computational indistinguishability, which underlines Definition 3.1, covers all possible efficient applications postulating that for all of them pseudorandom sequences are as good as truly random ones.

3.3.2 Some Basic Observations

We now present some basic observations regarding pseudorandom generators and the underlying notion of computational indistinguishability.

Amplifying the stretch function. Pseudorandom generators of any given stretch function, and in particular $\ell_1(k) \overset{\text{def}}{=} k + 1$, are easily converted into pseudorandom generators of any desired (polynomially bounded) stretch function, ℓ. Thus, when talking about the existence of pseudorandom generators, we may ignore the stretch function.

Construction 3.4 [184]: *Let G_1 be a pseudorandom generator with stretching function $\ell_1(k) = k + 1$, and ℓ be any polynomially bounded stretch function, which is polynomial-time computable. Let*

$$G(s) \overset{\text{def}}{=} \sigma_1 \sigma_2 \cdots \sigma_{\ell(|s|)},$$

where $x_0 = s$ and $x_i \sigma_i = G_1(x_{i-1})$, for $i = 1, ..., \ell(|s|)$. (That is, σ_i is the last bit of $G_1(x_{i-1})$ and x_i is the $|s|$-bit long prefix of $G_1(x_{i-1})$.)

Proposition 3.5 *G as defined in Construction 3.4 constitutes a pseudorandom generator.*

Proof Sketch: The proposition is proven using the *hybrid technique* (cf., [170, Sec. 3.2.3]): One considers distributions H_k^i (for $i = 0, ..., \ell(k)$) defined by $U_i^{(1)} P_{\ell(k)-i}(U_k^{(2)})$, where $U_i^{(1)}$ and $U_k^{(2)}$ are independent uniform distributions (over $\{0,1\}^i$ and $\{0,1\}^k$, respectively), and $P_j(x)$ denotes the j-bit long prefix of $G(x)$. The extreme hybrids correspond to $G(U_k)$ and $U_{\ell(k)}$, whereas distinguishability of neighboring hybrids can be worked into distinguishability of $G_1(U_k)$ and U_{k+1}. Loosely speaking, suppose one could distinguish H_k^i from H_k^{i+1}. Then, defining $f(s)$ (resp., $b(s)$) as the first $|s|$ bits (resp., last bit) of $G_1(s)$, and using $P_j(s) = b(s)P_{j-1}(f(s))$ (for $j \geq 1$), this means that one can distinguish $H_k^i \equiv (U_i^{(1)}, b(U_k^{(2)}), P_{(\ell(k)-i)-1}(f(U_k^{(2)})))$ from $H_k^{i+1} \equiv (U_i^{(1)}, U_1^{(1')}, P_{\ell(k)-(i+1)}(U_k^{(2')}))$. Incorporating the generation of $U_i^{(1)}$ and the evaluation of $P_{\ell(k)-i-1}$ into the distinguisher, one could distinguish $(f(U_k^{(2)}), b(U_k^{(2)})) \equiv G_1(U_k)$ from $(U_k^{(2')}, U_1^{(1')}) \equiv U_{k+1}$, in contradiction to the pseudorandomness of G_1. (For details see [170, Sec. 3.3.3].) $\square$

Derandomization of $\mathcal{BPP}$. Assuming the existence of pseudorandom generators and given Construction 3.4 and the above discussion, it follows that, for any constant $\epsilon > 0$, the randomness complexity of any polynomial-time algorithm (as a function of the input length n) can be shrinked to n^ϵ, without incurring any noticeable difference in its behavior. In particular, assuming that the original algorithm is a decision procedure for some language (in $\mathcal{BPP}$), then it is *infeasible to find* a (long enough) input on which the modified algorithm decides differently than the original one (e.g., the original algorithm accepts the input with probability at least $2/3$, whereas the modified algorithm accepts it with probability less than 0.6). However, this does not mean that such inputs do not exist (rather than being hard to find). Thus, in order to "derandomize" $\mathcal{BPP}$ we need a stronger notion of a pseudorandom generator; that is, one which can fool all polynomial-size circuits (and not merely all polynomial-time algorithms).

Definition 3.6 (strong pseudorandom generator – fooling circuits): *A deterministic polynomial-time algorithm G is called a* non-uniformly strong pseudorandom generator *if there exists a stretching function, $\ell : \mathbb{N} \mapsto \mathbb{N}$, so that for any family $\{C_k\}_{k \in \mathbb{N}}$ of polynomial-size circuits, for any positive polynomial p, and for all sufficiently large k's*

$$|\Pr[C_k(G(U_k)) = 1] - \Pr[C_k(U_{\ell(k)}) = 1]| < \frac{1}{p(k)}$$

Theorem 3.7 (Derandomization of $\mathcal{BPP}$ [351]): *If there exists non-uniformly strong pseudorandom generators then $\mathcal{BPP}$ is contained in $\cup_{\epsilon>0}\mathrm{Dtime}(t_\epsilon)$, where $t_\epsilon(n) \overset{\text{def}}{=} 2^{n^\epsilon}$.*

Proof Sketch: Given any $L \in \mathcal{BPP}$ and any $\epsilon > 0$, we let A denote the decision procedure for L and G denote a pseudorandom generator stretching n^ϵ-bit long seeds into $\mathrm{poly}(n)$-long sequences (to be used by A on input length n). We thus obtain an algorithm $A' = A_G$ (as in Construction 3.2). We note that A and A' may differ in their decision on at most finitely many inputs (or else we can incorporate such inputs, together with A, into a family of polynomial-size circuits which distinguishes $G(U_{n^\epsilon})$ from $U_{\mathrm{poly}(n)}$). Incorporating these finitely many inputs into A', and more importantly – emulating A' on each of the 2^{n^ϵ} possible random choices (i.e., seeds to G), we obtain a deterministic algorithm A'' as required. $\square$

We comment that stronger results regarding derandomization of $\mathcal{BPP}$ are presented in Section 3.4.

Computational Indistinguishability under multiple samples. The definition of computational indistinguishability underlying Definition 3.1 refers to distinguishers which obtain a single sample from each of the possible probability ensembles (i.e., $\{U_{\ell(k)}\}_{k\in\mathbb{N}}$ and $\{G(U_k)\}_{k\in\mathbb{N}}$). A more general definition refers to distinguishers which obtain several independent samples from each of the possible ensembles.[3]

Definition 3.8 (indistinguishability by multiple samples): *Let* $s : \mathbb{N} \mapsto \mathbb{N}$ *be polynomially-bounded. Two probability ensembles,* $X \stackrel{\text{def}}{=} \{X_k\}_{k\in\mathbb{N}}$ *and* $Y \stackrel{\text{def}}{=} \{Y_k\}_{k\in\mathbb{N}}$, *are* computationally indistinguishable by $s(\cdot)$ samples *if for every probabilistic polynomial-time algorithm,* D, *every polynomial* $p(\cdot)$, *and all sufficiently large* k's

$$\left| \Pr\left[D(X_k^{(1)}, ..., X_k^{(s(k))}) = 1\right] - \Pr\left[D(Y_k^{(1)}, ..., Y_k^{(s(k))}) = 1\right] \right| < \frac{1}{p(k)}$$

where $X_k^{(1)}$ *through* $X_k^{(s(k))}$ *and* $Y_k^{(1)}$ *through* $Y_k^{(s(k))}$ *are independent random variables, with each* $X_k^{(i)}$ *identical to* X_k *and each* $Y_k^{(i)}$ *identical to* Y_k.

Using the hybrid technique one can easily show that if both X and Y are polynomial-time constructible then computational indistinguishability by a single sample implies computational indistinguishability by any polynomial number of samples. (The ensemble $\{Z_k\}_{k\in\mathbb{N}}$ is said to be polynomial-time constructible if there exists a polynomial-time algorithm S so that $S(1^k)$ and Z_k are identically distributed.) The condition (of both ensembles being polynomial-time constructible) is essential; see [183, 193].

Non-triviality of Computational Indistinguishability. Clearly, any two distributions ensembles which are statistically close[4] are computationally indistinguishable. As noted above, there exist probability ensembles which are statistically far apart and yet are computationally indistinguishable [351, 180]. However, at least one of the probability ensembles in these results is *not* polynomial-time constructible. As we shall see below, the existence of one-way functions implies the existence of *polynomial-time constructible* probability ensembles which are statistically far apart and yet are computationally indistinguishable [211]. This sufficient condition is also necessary (cf., [166]).

---

[3] We have implicitly used the notion of a probability ensemble so far without explicitly defining it. As our usage of this term at this point is explicit, we now define it: By a **probability ensemble**, $\{Z_k\}_{k\in\mathbb{N}}$, we mean an infinite sequence of random variables such that each Z_k ranges over strings of length bounded by a polynomial in k.

[4] Two probability ensembles, $\{X_k\}_{k\in\mathbb{N}}$ and $\{Y_k\}_{k\in\mathbb{N}}$, are said to be statistically close if for every positive polynomial p and sufficient large k the variation distance between X_k and Y_k (i.e., $\frac{1}{2}\sum_z |\Pr[X_k = z] - \Pr[Y_k = z]|$) is bounded above by $1/p(k)$.

3.3.3 Constructions

The constructions surveyed in this section transform computation difficulty, in the form of one-way functions, into generators of pseudorandomness. Loosely speaking, a *polynomial-time computable* function is called one-way if any efficient algorithm can invert it only with negligible success probability. For simplicity we consider throughout this section only length-preserving one-way functions.

Definition 3.9 (one-way function): *A* one-way function, *f, is a polynomial-time computable function such that for every probabilistic polynomial-time algorithm A', every positive polynomial $p(\cdot)$, and all sufficiently large k's*

$$\Pr\left[A'(f(U_k)) \in f^{-1}(f(U_k))\right] < \frac{1}{p(k)}$$

We stress that both occurrences of U_k refer to the same random variable. That is, the above asserts that

$$\sum_{x \in \{0,1\}^k} 2^{-k} \cdot \Pr\left[A'(f(x)) \in f^{-1}(f(x))\right] < \frac{1}{p(k)}$$

Popular candidates for one-way functions are based on the conjectured intractability of Integer Factorization (cf., [288] for state of the art), the Discrete Logarithm Problem (cf., [289] analogously), and decoding of random linear code [181]. The infeasibility of inverting f yields a weak notion of unpredictability: For every probabilistic polynomial-time algorithm A (and sufficiently large k), it must be the case that $\Pr_i[A(i, f(U_k)) \neq b_i(U_k)] > 1/2k$, where the probability is taken uniformly over $i \in \{1, ..., k\}$ (and U_k), and $b_i(x)$ denotes the ith bit of x. A stronger (and in fact strongest possible) notion of unpredictability is that of a hard-core predicate. Loosely speaking, a *polynomial-time computable* predicate b is called a hard-core of a function f if all efficient algorithm, given $f(x)$, can guess $b(x)$ only with success probability which is negligible better than half.

Definition 3.10 (hard-core predicate [71]): *A polynomial-time computable predicate $b : \{0,1\}^* \mapsto \{0,1\}$ is called a* hard-core *of a function f if for every probabilistic polynomial-time algorithm A', every polynomial $p(\cdot)$, and all sufficiently large k's*

$$\Pr\left(A'(f(U_k)) = b(U_k)\right) < \frac{1}{2} + \frac{1}{p(k)}$$

Clearly, if b is a hard-core of a 1-1 polynomial-time computable function f then f must be one-way.[5] It turns out that any one-way function can be slightly modified so that it has a hard-core predicate.

---

[5] Functions which are not 1-1 may have hard-core predicates of information theoretic nature; but these are of no use to us here. For example, for $\sigma \in \{0,1\}$, $f(\sigma, x) = 0f'(x)$ has an "information theoretic" hard-core predicate $b(\sigma, x) = \sigma$.

Theorem 3.11 (A generic hard-core [182]): *Let f be an arbitrary one-way function, and let g be defined by $g(x, r) \stackrel{\text{def}}{=} (f(x), r)$, where $|x| = |r|$. Let $b(x, r)$ denote the inner-product mod 2 of the binary vectors x and r. Then the predicate b is a hard-core of the function g.*

A proof is presented in Appendix C.2. Finally, we get to the construction of pseudorandom generators.

Proposition 3.12 (A simple construction of pseudorandom generators): *Let b be a hard-core predicate of a polynomial-time computable 1-1 function f. Then, $G(s) \stackrel{\text{def}}{=} f(s)b(s)$ is a pseudorandom generator.*

Proof Sketch: Clearly the $|s|$-bit long prefix of $G(s)$ is uniformly distributed (since f is 1-1 and onto $\{0,1\}^{|s|}$). Hence, the proof boils down to showing that distinguishing $f(s)b(s)$ from $f(s)\sigma$, where σ is a random bit, yields contradiction to the hypothesis that b is a hard-core of f (i.e., that $b(s)$ is *unpredictable* from $f(s)$). Intuitively, such a distinguisher also distinguishes $f(s)b(s)$ from $f(s)\overline{b(s)}$, where $\overline{\sigma} = 1 - \sigma$, and so yields an algorithm for predicting $b(s)$ based on $f(s)$.   □

In a sense, the key point in the above proof is showing that the (obvious by definition) unpredictablity of the output of G implies its pseudorandomness. The fact that (next bit) unpredictability and pseudorandomness are equivalent in general is proven explicitly in the alternative presentation below.

An alternative presentation. Our presentation of the construction of pseudorandom generators, via Construction 3.4 and Proposition 3.12, is analogous to the original construction of pseudorandom generators suggested by by Blum and Micali [71]: Given an arbitrary stretch function $\ell : \mathbb{N} \mapsto \mathbb{N}$, a 1-1 one-way function f with a hard-core b, one defines

$$G(s) \stackrel{\text{def}}{=} b(x_1)b(x_2) \cdots b(x_{\ell(|s|)}),$$

where $x_0 = s$ and $x_i = f(x_{i-1})$ for $i = 1, ..., \ell(|s|)$. The pseudorandomness of G is established in two steps, using the notion of (next bit) unpredictability. An ensemble $\{Z_k\}_{k \in \mathbb{N}}$ is called **unpredictable** if any probabilistic polynoimal-time machine obtaining a prefix of Z_k fails to predict the next bit of Z_k with probability non-negligiblly higher than $1/2$.

1. One first proves that the ensemble $\{G(U_k)\}_{k \in \mathbb{N}}$, where U_k is uniform over $\{0,1\}^k$, is (next-bit) unpredictable (from right to left) [71].
   Loosely speaking, if one can predict $b(x_i)$ from $b(x_{i+1}) \cdots b(x_{\ell(|s|)})$ then one can predict $b(x_i)$ given $f(x_i)$ (i.e., by computing $x_{i+1}, ..., x_{\ell(|s|)}$ and so obtaining $b(x_{i+1}) \cdots b(x_{\ell(|s|)})$), in contradiction to the hard-core hypothesis.
2. Next, one uses Yao's observation by which a (polynomial-time constructible) ensemble is *pseudorandom if and only if it is* (next-bit) *unpredictable* (cf., [170, Sec. 3.3.4]).

Clearly, if one can predict the next bit in an ensemble then one can certainly distinguish this ensemble from the uniform ensemble (which in unpredictable regardless of computing power). However, here we need the other direction which is less obvious. Still, using a hybrid argument, one can show that (next bit) unpredictability implies indistinguishability from the uniform ensemble. Specifically, the ith hybrid takes the first i bits from the questionable ensemble and the rest from the uniform one. Thus, distinguishing the extreme hybrids implies distinguishing some neighboring hybrids, which in turn implies next-bit predictability.

A general condition for the existence of pseudorandom generators. Recall that given any one-way 1-1 function, we can easily construct a pseudorandom generator. Actually, the 1-1 requirement may be dropped, but the currently known construction – for the general case – is quite complex.

Theorem 3.13 (On the existence of pseudorandom generators [211]): *Pseudorandom generators exist if and only if one-way functions exist.*

To show that the existence of pseudorandom generators imply the existence of one-way functions, consider a pseudorandom generator G with stretch function $\ell(k) = 2k$. For $x, y \in \{0,1\}^k$, define $f(x, y) \stackrel{\text{def}}{=} G(x)$, and so f is polynomial-time computable (and length-preserving). It must be that f is one-way, or else one can distinguish $G(U_k)$ from U_{2k} by trying to invert and checking the result: Inverting f on its range distribution refers to the distribution $G(U_k)$, whereas the probability that U_{2k} has inverse under f is negligible.

The interesting direction is the construction of pseudorandom generators based on any one-way function. In general (when f may not be 1-1) the ensemble $f(U_k)$ may not be pseudorandom, and so Construction 3.12 (i.e., $G(s) = f(s)b(s)$, where b is a hard-core of f) cannot be used *directly*. One idea of [211] is to hash $f(U_k)$ to an almost uniform string of length related to its entropy, using Universal Hash Functions [92]. (This is done after guaranteeing, that the logarithm of the probability mass of a value of $f(U_k)$ is typically close to the entropy of $f(U_k)$.)[6] But "hashing $f(U_k)$ down to length comparable to the entropy" means shrinking the length of the output to, say, $k' < k$. This foils the entire point of stretching the k-bit seed. Thus, a second idea of [211] is to compensate for the $k - k'$ loss by extracting these many bits from the seed U_k itself. This is done by hashing U_k, and the point is that the $(k - k' + 1)$-bit long hash value does not make the inverting task any easier. Implementing these ideas turns out to be more difficult than it seems, and indeed an alternative construction would be most appreciated.

---

[6] Specifically, given an arbitrary one way function f', one first constructs f by taking a "direct product" of sufficiently many copies of f'. For example, for $x_1, ..., x_{k^2} \in \{0,1\}^k$, we let $f(x_1, ..., x_{k^2}) \stackrel{\text{def}}{=} f'(x_1), ..., f'(x_{k^2})$.

On constructing non-uniformly strong pseudorandom generators.
Non-uniformly strong pseudorandom generators (i.e., which produce sequences indistinguishable by polynomial-size circuits as in Definition 3.6) can be constructed analogously using any one-way function which is hard to invert by any non-uniform family of polynomial-size circuits (rather than by probabilistic polynomial-time machines). In fact, the construction can be simplified in this case (cf., [217]).

Advanced comment regarding other strong notions (of pseudorandom generators): An alternative strengthening of Definition 3.1 amounts to explicitly quantifying the resources and success gaps of distinguishers. These quantities will be bounded as a function of the seed length (i.e., k) rather as a function of the sequence which is being examined (i.e., $\ell(k)$). For a class of time bounds $\mathcal{T}$ (e.g., $\mathcal{T} \stackrel{\text{def}}{=} \{t(k) \stackrel{\text{def}}{=} 2^{c\sqrt{k}}\}_{c\in\mathbb{N}}$) and a class of noticeable functions (e.g., $\mathcal{F} = \{f(k) \stackrel{\text{def}}{=} 1/t(k) : t \in \mathcal{T}\}$), we say that a pseudorandom generator, G, is $(\mathcal{T}, \mathcal{F})$-**strong** if for any probabilistic algorithm D having running-time bounded by a function in $\mathcal{T}$ (applied to k)[7], for any function f in $\mathcal{F}$, and for all sufficiently large k's

$$| \Pr[D(G(U_k)) = 1] - \Pr[D(U_{\ell(k)}) = 1]| < f(k)$$

An analogous strengthening may be applied to the definition of one-way functions. Doing so reveals the weakness of the result in [211]: It only implies that for some $\epsilon > 0$ ($\epsilon = 1/5$ will do), for any $\mathcal{T}$ and $\mathcal{F}$, the existence of $(\mathcal{T}, \mathcal{F})$-strong one-way functions implies the existence of $(\mathcal{T}', \mathcal{F}')$-strong pseudorandom generators, where $\mathcal{T}' = \{t'(k) \stackrel{\text{def}}{=} t(k^\epsilon)/\text{poly}(k) : t \in \mathcal{T}\}$ and $\mathcal{F}' = \{f'(k) \stackrel{\text{def}}{=} \text{poly}(k) \cdot f(k^\epsilon) : f \in \mathcal{F}\}$. What we *would like* to have is an analogous result with $\mathcal{T}' = \{t'(k) \stackrel{\text{def}}{=} t(k)/\text{poly}(k) : t \in \mathcal{T}\}$ and $\mathcal{F}' = \{f'(k) \stackrel{\text{def}}{=} \text{poly}(k) \cdot f(k) : f \in \mathcal{F}\}$.

3.3.4 Pseudorandom Functions

Pseudorandom generators allow to efficiently generate long pseudorandom sequences from short random seeds. Pseudorandom functions (defined below) are even more powerful: They allow efficient direct access to a huge pseudorandom sequence (which is not even feasible to scan bit-by-bit). Put in other words, pseudorandom functions can replace truly random functions in any efficient application (e.g., most notably in cryptography).

Definition 3.14 (pseudorandom functions [174]): *A* pseudorandom function (ensemble)*, with length parameters* $\ell_D, \ell_R : \mathbb{N} \mapsto \mathbb{N}$*, is a collection of functions* $\{f_s : \{0,1\}^{\ell_D(|s|)} \mapsto \{0,1\}^{\ell_R(|s|)}\}_{s\in\{0,1\}^*}$ *satisfying*

---

[7] That is, when examining a sequence of length $\ell(k)$ algorithm D makes at most $t(k)$ steps, where $t \in \mathcal{T}$.

- (efficient evaluation): *There exists an efficient* (deterministic) *algorithm which given a seed, s, and an $\ell_D(|s|)$-bit argument, x, returns the $\ell_R(|s|)$-bit long value $f_s(x)$.*
- (pseudorandomness): *For every probabilistic polynomial-time oracle machine, M, for every positive polynomial p and all sufficiently large k's*

$$\left| \Pr[M^{f_{U_k}}(1^k) = 1] - \Pr[M^{F_k}(1^k) = 1] \right| < \frac{1}{p(k)}$$

where F_k denotes a uniformly selected function mapping $\{0,1\}^{\ell_D(k)}$ to $\{0,1\}^{\ell_R(k)}$.

Suppose, for simplicity, that $\ell_D(k) = k$ and $\ell_R(k) = 1$. Then a function uniformly selected among 2^k functions (of a pseudorandom ensemble) presents an input-output behavior which is computationally indistinguishable from the one of a function selected at random among all the 2^{2^k} Boolean functions. Contrast this with the 2^k pseudorandom sequences, produced by a pseudorandom generator, which are computationally indistinguishable from a sequence selected uniformly among all the $2^{\text{poly}(k)}$ many sequences. Still pseudorandom functions can be constructed from any pseudorandom generator.

Theorem 3.15 (How to construct pseudorandom functions [174]): *Let G be a pseudorandom generator with stretching function $\ell(k) = 2k$, let $G_0(s)$ (resp., $G_1(s)$) denote the first (resp., last) $|s|$ bits in $G(s)$, and*

$$G_{\sigma_{|s|} \cdots \sigma_2 \sigma_1}(s) \stackrel{\text{def}}{=} G_{\sigma_{|s|}}(\cdots G_{\sigma_2}(G_{\sigma_1}(s)) \cdots)$$

Then, the function ensemble $\{f_s : \{0,1\}^{|s|} \mapsto \{0,1\}^{|s|}\}_{s \in \{0,1\}^*}$, where $f_s(x) \stackrel{\text{def}}{=} G_x(s)$, is pseudorandom with length parameters $\ell_D(k) = \ell_R(k) = k$.

The above construction can be easily adapted to any (polynomially-bounded) length parameters $\ell_D, \ell_R : \mathbb{N} \mapsto \mathbb{N}$.

Proof Sketch: The proof uses the hybrid technique: The ith hybrid, H_k^i, is a function ensemble consisting of $2^{2^i \cdot k}$ functions $\{0,1\}^k \mapsto \{0,1\}^k$, each defined by 2^i random k-bit strings, denoted $\langle s_\alpha \rangle_{\alpha \in \{0,1\}^i}$. The value of such function at $x = \beta\alpha$, with $|\alpha| = i$, is $G_\beta(s_\alpha)$. The extreme hybrids correspond to our indistinguishability claim (i.e., $H_k^0 \equiv f_{U_k}$ and $H_k^k \equiv F_k$), and neighboring hybrids correspond to our indistinguishability hypothesis (specifically, to the indistinguishability of $G(U_k)$ and U_{2k} under multiple samples).   □

We mention that pseudorandom functions have been used to derive negative results in computational learning theory [342] and in complexity theory (cf., Natural Proofs [311]).

3.4 Derandomization of Time-complexity Classes

Recall the proof of Theorem 3.7: A pseudorandom generator was used to shrink the randomness complexity of a BPP-algorithm, and derandomization was achieved by scanning all possible seeds to the generator. A key observation of [281, 286] is that whenever a pseudorandom generator is used this way, there is no point in insisting that it runs in time polynomial in its seed length. Instead, it suffices to require that the generator runs in time exponential in its seed length (as we are incurring such a time factor anyhow due to the scanning of all possible seeds). Thus, the generator may have running-time greater than the distinguisher it is designed to fool. This observation has opened the door to a sequence of derandomization results [286, 26, 216, 221] culminating in the following theorem, where $\mathcal{E} \stackrel{\text{def}}{=} \cup_c \text{Dtime}(t_c)$ with $t_c(n) = 2^{cn}$.

Theorem 3.16 (Derandomization of BPP, revisited [221]): *Suppose that there exists a language $L \in \mathcal{E}$ having almost-everywhere exponential circuit complexity* (i.e., there exists a constant $b > 0$ such that, for all but finitely many k's, any circuit C_k which correctly decides L on $\{0,1\}^k$ has size at least 2^{bk}). *Then, $\mathcal{BPP} = \mathcal{P}$.*

Proof Sketch: Underlying the proof is a construction of a pseudorandom generator due to Nisan and Wigderson [281, 286]. This construction utilizes a predicate computable in exponential-time but unpredictable, even to within a particular exponential advantage, by any circuit family of a particular exponential size. (The crux of [221] is supplying such a predicate, given the hypothesis; their argument utilizes [281, 26, 182, 4, 216].) Given such a predicate the generator works by evaluating the predicate on exponentially-many subsequences of the bits of the seed so that the intersection of any two subsets is relatively small.[8] Thus, for some constant $b > 0$ and all k's, the generator stretches seeds of length k into sequences of length 2^{bk} which (as loosely argued below) cannot be distinguished from truly random sequences by any circuit of size 2^{bk}.[9] The derandomization of $\mathcal{BPP}$ proceeds by setting the seed-length to be logarithmic in the input length, and utilizing the above generator.

The above generator fools circuits of the stated size, even when these circuits are presented with the seed as auxiliary input. (These circuits are smaller than the running time of the generator and so they cannot just evaluate the generator on the given seed.) The proof that the generator fools

---

[8] These subsets have size linear in the length of the seed, and intersect on a constant fraction of their respective size. Furthermore, they can be determined within exponential-time.

[9] Thus, this generator is only "moderately more complex" than the distinguisher: Viewed in terms of its output, the generator works in time polynomial in the length of the output, whereas the output fools circuits of size which is a (smaller) polynomial in the length of the output.

such circuits refers to the characterization of pseudorandom sequences as un-predictable ones. Thus, one proves that the next bit in the generator's output cannot be predicted given all previous bits (as well as the seed). Assuming that a small circuit can predict the next bit, we construct a circuit for pre-dicting the hard predicate. The new circuit incorporates the best (for such prediction) augmentation of the input to the circuit into a seed for the gene-rator (i.e., the bits not in the specific subset of the seed are fixed in the best way). The key observation is that all other bits in the output of the generator depend only on a small fraction of the input bits (i.e., recall the small in-tersection clause above), and so circuits for computing these other bits have relatively small size (and so can be incorporated in the new circuit). Using all these circuits, the new circuit forms the adequate input for the next-bit predicting circuit, and outputs whatever the latter circuit does.   □

Derandomization of constant-depth circuits. The same underlying idea, yet with a different setting of parameters and using the PARITY func-tion (which is hard for "small" constant-depth circuits [352, 207]), was used in the context of constant-depth circuits. The aim was to derandomize $\mathcal{RAC}_0$ (i.e., random $\mathcal{AC}_0$), or put in other words – given a constant-depth circuit to *deterministically* approximate (up-to an additive error) the fraction of inputs which evaluate to some output. The result obtained in [281] implies that, for any constant d, given a depth-d circuit C, one can approximate the fraction of the number of inputs to C which evaluate to 1 to within *additive error* 0.01 by a deterministic quasi-polynomial-time algorithm. For the special case of approximating the number of satisfying assignment of a DNF formula, *relative error* approximations can be obtained by employing the reduction of [229]. (See also improvements in [253].)

3.5 Space Pseudorandom Generators

In the previous two sections we have considered generators the output of which is indistinguishable by any efficient procedures. The latter were mo-deled by time-bounded computations; specifically, polynomial-time compu-tations. A finer characterization of time-bounded computations is obtained by considering the space-complexity. Unfortunately, natural notions of space-bounded computations are quite subtle – especially when randomization or non-determinism are concerned (cf., [317]). Two major issues are:

1. *Time bound*: Whether one restricts these machines to have time-complex-ity at most exponential in the space-complexity (like in the deterministic case).[10] Indeed, following most work in the area, we do postulate so.

---

[10] Alternatively, one can ask whether these machines must always halt or only halt with probability approaching 1. It can be shown that the only way to ensure "absolute halting" is to have the time-complexity at most exponential in the space-complexity [317].

2. *Access to random tape*: Whether the space-bounded machine has one-way or two-way access to the randomness tape. (Allowing two-way access means that the randomness is recorded for free; that is, without being accounted for in the space-bound.) An alternative presentation of the question refers to whether the randomness is to be considered as taking place on-line or whether it is done off-line and given as auxiliary input (to which one has a two-way access). Again, following most work in the area, we consider one-way access.[11]

In accordance with the resulting definition of randomized bounded-space computation, we consider space-bounded distinguishers which have a one-way access to the input sequence which they examine. As all known constructions remain valid also when these distinguishers are non-uniform, we define this stronger notion below. In such cases one may assume, without loss of generality, that the running-time of the distinguisher equals the length of its input (i.e., the inspected sequence). A non-uniform machine of space $s : \mathbb{N} \mapsto \mathbb{N}$ is thus a family, $\{D_k\}_{k \in \mathbb{N}}$, of directed layered graphs so that D_k has at most $2^{s(k)}$ vertices at each layer, and labeled directed edges from each layer to the next layer.[12] Each vertex has two (possibly parallel) outgoing directed edges, one labeled 0 and the other labeled 1. Such a non-uniform machine yields a natural notion of decision (i.e., consider a fixed partition of the vertices of the last layer, and define the result of the computation according to the vertex reached when following the path labeled correspondingly to the input).

Definition 3.17 (Indistinguishability w.r.t space-bounded machines):

- *For a non-uniform machine, $\{D_k\}_{k \in \mathbb{N}}$, and two probability ensembles, $\{X_k\}_{k \in \mathbb{N}}$ and $\{Y_k\}_{k \in \mathbb{N}}$, the function $d : \mathbb{N} \mapsto [0,1]$ defined as $d(k) \overset{\text{def}}{=} |\Pr[D_k(X_k) = 1] - \Pr[D_k(Y_k) = 1]|$ is called the* distinguishability-gap *of $\{D_k\}$ between the two ensembles.*
- *A probability ensemble, $\{X_k\}_{k \in \mathbb{N}}$, is called (s, ϵ)-*pseudorandom *if for any (non-uniform) $s(\cdot)$-space-bounded machine, the distinguishability-gap of the machine between $\{X_k\}_{k \in \mathbb{N}}$ and a uniform ensemble (of the same length) is at most $\epsilon(\cdot)$.*
- *A deterministic algorithm G of stretch function ℓ is called a (s, ϵ)-*pseudorandom generator *if the ensemble $\{G(U_k)\}_{k \in \mathbb{N}}$ is (s, ϵ)-pseudorandom, where U_k denotes the uniform distribution over $\{0,1\}^k$.*

---

[11] We note that the fact that we restrict our attention to one-way access is instrumental in obtaining space-robust generators without making intractability assumptions. Analogous generators for two-way space-bounded computations would imply hardness results of a breakthrough nature in the area.

[12] Note that the space bound of the machine is stated in terms of a parameter k, rather than in terms of the length of its input. In the sequel this parameter will be set to the length of a seed to a pseudorandom generator. We warn that our presentation here is indeed non-standard for this area. To compensate for this, we will also state the consequences in the standard format.

Following are the two major results regarding pseudorandom generators with respect to space-bounded machines. In contrast to the pseudorandom generators in the previous two sections, the existence of the "bounded-space resilient pseudorandom generators" does not depend on any computational assumptions.

Theorem 3.18 (Nisan's Generator [282]): *For every $s : \mathbb{N} \mapsto \mathbb{N}$, there exists a $(s, 2^{-s})$-pseudorandom generator of stretch function $\ell(k) = 2^{k/O(s(k))}$. The generator works in space linear in the length of the seed, and in time linear in the stretch function.*

In other words, we have a generator which takes a random seed of length $k = O(t \cdot m)$ and produce sequences of length 2^t which look random to any m-space-bounded machine. In particular, using a random seed of length $k = O(m^2)$, one can produce sequences of length 2^m which look random to any m-space bounded machine. Thus, one may replace random sequences used by ANY space-bounded computation, by sequences which are efficiently generated from random seeds of length quadratic in the space bound. (The common instantiation is for log-space machines.)

Theorem 3.19 (The Nisan-Zuckerman Generator [287]): *For any polynomial p, there exists a function $s(k) = k/O(1)$ and a $(s, 2^{-\sqrt{s}})$-pseudorandom generator of stretch function p. The generator works in linear-space and polynomial-time* (both stated in terms of the length of the seed).

In other words, we have a generator which takes a random seed of length $k = O(m)$ and produce sequences of length $\text{poly}(m)$ which look random to any m-space-bounded machine. Thus, one may *convert* ANY *randomized computation utilizing polynomial-time and linear-space into a functionally equivalent randomized computation of similar time and space complexities which uses only a linear number of coin tosses.* (The above two results have been "interpolated" in [17]: There exists a parameterized family of space pseudorandom generators which includes both the above as extreme special cases.)

Comments on the proofs of the above two theorems. In both cases, we describe the construction by starting with an adequate distinguisher and showing how the input distribution it examines can be modified (from the uniform one into a pseudorandom one) without the distinguisher noticing the difference.

Theorem 3.18 is proven by using the "mixing property" of Universal Hash Functions [92]. A family of functions H_n which map $\{0,1\}^n$ to itself is called *mixing* if for every pair of subsets $A, B \subseteq \{0,1\}^n$ for all but few of the functions $h \in H_n$,

$$\Pr[U_n \in A \wedge h(U_n) \in B] \approx \frac{|A|}{2^n} \cdot \frac{|B|}{2^n}$$

Given a $s(k)$-space distinguisher D_k as above, we set $n \stackrel{\text{def}}{=} O(s(k))$ and $\ell' \stackrel{\text{def}}{=} \ell(k)/n$, and consider an auxiliary "distinguisher" D_k' which is a directed layered graph with ℓ' layers and $2^{s(k)}$ vertices in each layer. Each vertex has directed edges going to each vertex of the next layer and these edges are labelled with (possiblly empty) subsets of $\{0,1\}^n$, where these subsets form a partition of $\{0,1\}^n$. The graph D_k' simulates D_k in the obvious manner (i.e., the computation of D_k' on input of length $\ell(k) = \ell' \cdot n$ is defined by breaking the input into consequetive blocks of length n and following the path of edges labelled by the subsets contaning the corresponding block). For each pair of neighboring vertices, u and v (in layers i and $i+1$, respectively), consider the label, $L_{u,v} \subseteq \{0,1\}^n$, of the edge going from u to v. Similarly, for a vertex w at layer $i+2$, we consider the label $L_{v,w}'$ of the edge from v to w. By the above mixing property, for all but a few of $h \in H_n$,

$$\Pr[U_n \in L_{u,v} \wedge h(U_n) \in L_{v,w}'] \approx \Pr[U_n \in L_{u,v}] \cdot \Pr[U_n \in L_{v,w}']$$

Thus, replacing the coins in the second block (i.e., used in transitions from layer $i+1$ to layer $i+2$) by the value of h applied to the outcomes of the coins used in the first block (i.e., in transitions from layer i to $i+1$), approximately maintains the probability that D_k' moves from u to w via v. The same (with "few" being $2^{3s(k)} \cdot \ell'$ times larger here) holds for every triple of vertices in any three layers as above. The point is that we can use the same h in all these approximations. Thus, at the cost of extra $|h|$ random bits, we can reduce the number of true random coins used in transitions on D_k' by a factor of 2, without significantly effecting its decision. In other words, at the cost of extra $|h|$ random bits, we can effectively contract the distinguisher to half its length.[13] Repeating the process for a logarithmic (in D_k''s length) number of times we obtain a distinguisher which only examines n bits at which point we stop. In total we have used $\log_2(\ell(k)/O(s(k)))$ random hash functions, which means that we can generate a sequence which fools the original D_k using a seed of length $n + \log_2 \ell(k) \cdot |h|$, which for adequate family H_n yields the claimed seed length of $O(s(k) \cdot \log_2 \ell(k)) = k$.

Theorem 3.19 is proven by using a much more powerful tool – the extractor (as defined in Section 3.6). The basic idea is that when D_k is at some distant layer, say at layer t, it typically "knows" little about the random choices which led it there: That is, it has only $s(k)$ bits of memory which leaves out $t - s(k)$ bits of uncertainty (or randomness). Thus much of the randomness which led D_k to its current state may be "re-used" (or "recycled"). To reuse these bits we need to extract *almost* uniform distribution on strings of length, say, $t - s(k) - o(k)$ out of a distribution which has entropy $t - s(k)$ (actually a stronger technical condition need and can be imposed on the

---

[13] That is, fixing a good h as above, we can replace the 2-paths in D_k' by edges in a new distinguisher D_k'', so that r is in the set labeling an edge u–w in D_k'' iff for some v, the string r is in the label of the edge u–v in D_k' and $h(r)$ is in the label of the edge v–w (also in D_k').

distribution). Furthermore, such an extraction requires some – yet relatively few – truly random bits. In particular, [287] used $\sqrt{k}/2$ bits towards this end (and the extracted bits are $\exp(-\sqrt{k})$ away from uniform). An important point is how to use the above argument repeatedly. We break the seed into two parts, $\rho \in \{0,1\}^{k/2}$ and $r_1, ..., r_{\sqrt{k}}$ where $|r_i| = \sqrt{k}/2$, and set $n = k/3$. Looking at layer $i \cdot n$, we consider the information known about ρ (rather the information known about the last n steps). Thus, using r_i, we can extract $(k/2) - s(k) - o(k) > k/3 = n$ almost-random bits required for the next n steps. Hence, using k random bits we were able to produce a sequence of length $\sqrt{k} \cdot n = k^{3/2}/3$ which fools machines of space bound, say, $s(k) = k/10$. Using sequential composition, one may amplify the stretch function up-to any polynomial p at the expense of fooling only k/c-space machines, where c depends p.

Derandomization of space-complexity classes: Utilizing the specific structure of Nisan's Generator led to showing that randomized log-space can be simulated in deterministic polynomial-time and polylogarithmic-space (i.e., $\mathcal{RL} \subseteq \mathcal{SC}$) [283]. Thus, $\mathcal{RL}$ (and actually $\mathcal{BPL}$) were placed in a class not known to contain $\mathcal{NL}$. Another such result was subsequently obtained in [319]: Randomized log-space can be simulated in deterministic space $o(\log^2)$; specifically, in space $\log^{3/2}$. A better simulation is currently known for the archetypical (but not known to be complete) problem of $\mathcal{RL}$; that is, undirected connectivity [6]. Specifically, by [16] (improving over [285]), the problem is solvable by a deterministic algorithm of space complexity $O(\log^{4/3} n)$, where n is the size of the graph.

3.6 Special Purpose Generators

In this section we consider even weaker types of pseudorandom generators, producing sequences which can fool only very restricted types of distinguishers. Still, such generators have many applications in complexity theory and in the design of algorithms.

Technically speaking, the material presented in this section is quite interleaved. Furthermore, some of it is related to the results presented in the previous two sections (e.g., Theorem 3.18 uses ideas implicit in the material below, whereas Theorems 3.16 and 3.19 rely explicitly on results discussed below).

Our choice is to start with the simplest of these generators – the pairwise-independent generator [103], and its generalization [9] to t-wise independence, for any $t \geq 2$. Such generators perfectly fool any distinguisher which only observe t fixed locations in the output sequence. This leads naturally to almost pairwise (or t-wise) independence generators, which also fool (but non-perfectly) such distinguishers. The latter generators are implied by a stronger class of generators which is of independent interest – the small-bias generators [274]. Small-bias generators fool any linear test (i.e., any distinguisher which merely considers the XOR of some fixed locations in the input sequence). We then turn to the Expander Random Walk Generator – this generator produces a sequence of strings which hit any dense subset of strings with probability which is close to the hitting probability of a truly random sequence. A generalization, called a *sampler*, generates a sequence of sample points from which one can approximate the average value of any fixed function (which maps strings into a bounded interval of reals). Finally, we consider the related notions of a *disperser* and an *extractor*.

Comment regarding our parameterization: To maintain consistency with prior sections, we continue to present the generators in terms of the seed length, denoted k. Since this is not the common presentation for most results presented below, we provide (in footnotes) the common presentation where the seed length is determined as a function of other parameters.

3.6.1 Pairwise-Independence Generators

A *t-wise independence generator* of block-size $b : \mathbb{N} \mapsto \mathbb{N}$ (and stretch function ℓ) is an efficient (e.g., works in time polynomial in the output length) deterministic algorithm which expands a k-bit long random seed into a sequence of $\ell(k)/b(k)$ strings, each of length $b(k)$, such that any t blocks are uniformly and independently distributed in $\{0,1\}^{t \cdot b(k)}$. In case $t = 2$, we call the generator *pairwise independent*.

Proposition 3.20 (*t-wise independence generator* [103, 9]):[14] *Let t be a fixed integer, and suppose that $b(k) = k/t$, $\ell'(k) = \ell(k)/b(k)$ and $\ell'(k) < 2^{b(k)}$. Associate both $\{0,1\}^{b(k)}$ and $\{1, 2, ..., 2^{b(k)}\}$ with the field $\mathrm{GF}(2^{b(k)})$, and let $\alpha_1, ..., \alpha_{\ell'(k)}$ be distinct non-zero elements of this field. For $s_0, s_1, ..., s_{t-1} \in \{0,1\}^{b(k)}$, let*

$$G(s_0, s_1, ..., s_{t-1}) \stackrel{\text{def}}{=} \left(\sum_{j=0}^{t-1} s_j \alpha_1^j , \sum_{j=0}^{t-1} s_j \alpha_2^j , ..., \sum_{j=0}^{t-1} s_j \alpha_{\ell'(k)}^j \right)$$

where the arithmetic is that of $\mathrm{GF}(2^{b(k)})$. Then, G is a t-wise independence generator of block-size b.

To make the above generator totally explicit, we need an explicit representation of $\mathrm{GF}(2^{b(k)})$, which requires an irreducible polynomial of degree $b(k)$ over $\mathrm{GF}(2)$. For specific values of $b(k)$ a good representation exists: Specifically, for $d \stackrel{\text{def}}{=} b(k) = 2 \cdot 3^e$ (with e integer), the polynomial $x^d + x^{d/2} + 1$

---

[14] The common parameterization of t-wise independence generator is as follows. Given parameters b and $\ell' < 2^b$, and a uniformly chosen seed of length $t \cdot b$, one can efficiently and deterministically generate a random sequence of ℓ' strings, each of length b, which are t-wise independent.

is irreducible over $GF(2)$ [204, p. 96]. Alternatively, for $t = 2$, one may use affine transformations defined by random Toeplitz matrices.[15] That is,

Proposition 3.21 (Alternative pairwise independence generator [92]):[16] *Let $\ell'(k) = \ell(k)/b(k)$ and $m(k) = \lceil \log_2 \ell'(k) \rceil$, and suppose that $k = 2b(k) + m(k) - 1$. Associate $\{0,1\}^n$ with the n-dimensional vector space over $GF(2)$, and let $v_1, ..., v_{\ell'(k)}$ be distinct vectors in the $m(k)$-dimensional vector space. For $s \in \{0,1\}^{b(k)+m(k)-1}$, $r \in \{0,1\}^{b(k)}$ and $i = 1, ..., \ell'(k)$, let*

$$G(s,r) \overset{\text{def}}{=} (T_s v_1 + r, \; T_s v_2 + r, \; ..., \; T_s v_{\ell'(k)} + r)$$

where T_s is an $b(k)$-by-$m(k)$ Toeplitz matrix specified by the string s, is a pairwise independence generator of block-size b.

Pairwise independence generators do suffice for a variety of applications (cf., [349, 254]). In particular, we mention the application to sampling discussed in Section 3.6.4, and the celebrated derandomization by Luby [250]. The latter uses the fact that the analysis of the target randomized algorithm only relies on the hypothesis that some objects are selected in pairwise independent manner. Thus, such weak generators do suffice to fool distinguishers which are derived from some natural and interesting algorithms.

We remark that for constant $t \geq 2$, the cost of derandomization (i.e., going over all 2^k possible seeds) can be made exponential in the block-size (i.e., $b(k) = O(k/t)$) and polynomial in the number of blocks (i.e., $\ell'(k) \leq 2^{b(k)} = \exp(k/t)$). (We stress that it is important to have the cost of derandomization be polynomial in the length of the produced pseudorandom sequence, since the latter is typically polynomially-related to the length of the input to the algorithm we wish to derandomize.) Thus, *whenever the analysis of a randomized algorithm can be based on a constant amount of independence* between (feasibly-many) random choices, each made inside a feasible domain, *a feasible derandomization is possible*. On the other hand, the relationship $\ell(k) = \exp(k/t)$ is the best possible (cf., [101]), and so one cannot produce from a seed of length k an $\exp(k/O(1))$-long sequence of non-constant independence. Technically speaking, t-wise independent generators of stretch ℓ require a seed of length $\Omega(t \cdot \log \ell)$. In the next subsection we will see that meaningful approximations may be obtained with much shorter seeds.

---

[15] A Toeplitz matrix is a matrix with all diagonals being homogeneous; that is, $T = (t_{i,j})$ is a Toeplitz matrix if $t_{i,j} = t_{i+1,j+1}$, for all i, j. Note that a Toeplitz matrix is determined by its first row and first column (i.e., the values of $t_{1,j}$'s and $t_{i,1}$'s).

[16] The common parameterization of this pairwise independence generator is as follows. Given parameters b and $\ell' \leq 2^b$, and a uniformly chosen seed of length $2b + \lceil \log_2 \ell' \rceil - 1$, one can efficiently and deterministically generate a random sequence of ℓ' strings, each of length b, which are pairwise independent.

3.6.2 Small-Bias Generators

Trying to go beyond constant-independence in derandomizations (as above) was the primary motivation of Naor and Naor [274], and is in fact an important application of the notion of small-bias generators. Let $\epsilon : \mathbb{N} \mapsto [0, 1]$. An ϵ-bias generators with stretch function ℓ is an efficient (e.g., polynomial in $\ell(k)$ time) deterministic algorithm which expands a k-bit long random seed into a sequence of $\ell(k)$ bits, so that for any fixed (non-empty) set $S \subseteq \{1, ..., \ell(k)\}$ the bias of the output sequence over S is at most $\epsilon(k)$, where the bias of a sequence of n (possibly dependent) Boolean random variables $\zeta_1, ..., \zeta_n \in \{0, 1\}^n$ over $S \subseteq \{1, .., n\}$ is defined as $2 \cdot |\Pr[\oplus_{i \in S} \zeta_i = 1] - 0.5|$.[17]

Theorem 3.22 (small-bias generators [274]):[18] *Let the functions ℓ and ϵ be so that $k = O(\log(\ell(k)/\epsilon(k)))$. Then, there exists an ϵ-bias generator with stretch function ℓ operating in time polynomial in $\ell(k)$.*

Three alternative simpler constructions (i.e., proofs of Theorem 3.22) are given in [11]. One of these is based on Linear Feedback Shift Registers. Loosely speaking, the first half of the seed, denoted $f_0 f_1 \cdots f_{(k/2)-1}$, is interpreted as a (non-degenerate) feedback rule[19], the other half, denoted $s_0 s_1 \cdots s_{(k/2)-1}$, is interpreted as "the start sequence" and the ouput sequence, denoted $r_0 r_1 \cdots r_{\ell(k)-1}$, is obtained by setting $r_i = s_i$ for $i < k/2$ and $r_i = \sum_{j=0}^{(k/2)-1} f_j \cdot r_{i-(k/2)+j}$ for $i \geq k/2$.

Small-bias generators have been used in a variety of areas (e.g., non-approximation [212], structural complexity [291], and applied cryptography [236]). In addition, they seem an important tool in the design of various types of "pseudorandom" objects; see below.

Approximate independence generators. As observed in [345], ϵ-bias is related to approximate limited independence. Actually, a restricted type of ϵ-bias – in which only subsets of size $t(k)$ are required to have bias bounded by ϵ implies that the variation distance (i.e., Norm-1 distance) of any $t(k)$ bits in

---

[17] The factor of 2 was introduced so to make these biases correspond to the Fourier coefficients of the distribution (viewed as a function from $\{0, 1\}^n$ to the reals). To see the correspondance one should replace $\{0, 1\}$ by $\{\pm 1\}$, and substitute XOR by multiplication. The bias with respect to set S is thus written as $\text{Exp}[\prod_{i \in S} \zeta_i]$ ($= \Pr[\prod_{i \in S} \zeta_i = +1] - \Pr[\prod_{i \in S} \zeta_i = -1]$), which is merely the Fourier coefficient corresponding to S.

[18] Here the common parameterization is merely a point of view: Rather than saying that the functions ℓ and ϵ satisfy $k = O(\log(\ell(k)/\epsilon(k)))$, one says that given desired parameters ℓ and ϵ one sets $k = O(\log(\ell/\epsilon))$. We also comment that using [11] the constant in the O-notation is merely 2 (i.e., $k \approx 2\log_2(\ell/\epsilon)$), whereas using [274] $k \approx \log_2 \ell + 4\log_2(1/\epsilon)$.

[19] That is, $f_0 = 1$ and $f(t) \stackrel{\text{def}}{=} t^{k/2} + \sum_{j=0}^{(k/2)-1} f_j \cdot t^j$ is an irreducible polynomial over $GF(2)$.

the sequence from the uniform distribution is at most $2^{t(k)/2} \cdot \epsilon(k)$. (The max-norm of the difference is bounded by $\epsilon(k)$.)[20] Combining Theorem 3.22, the above observation, and the linearity of the construction in Proposition 3.20, one obtains generators with $\exp(k)$ stretch function which are approximately $t(k)$-independent, for non-constant $t(k)$ (cf., [274]). Specifically, one may obtain generators with stretch function ℓ, producing sequences in which any $t(k)$ positions are at most $\epsilon(k)$-away from uniform (in variation distance), provided that $k = O(t(k)+\log(1/\epsilon(k))+\log\log \ell(k))$.[21] (In particular, we may have $\ell(k) = 2^{k/O(1)}$, $t(k) = O(\log \ell(k))$, and $\epsilon(k) = 2^{-O(t(k))}$.) Thus, whenever the analysis of a randomized algorithm can be based on a logarithmic amount of (almost) independence between feasibly-many Boolean random choices, a feasible derandomization is possible. Extensions to non-Boolean choices are considered in [128, 95, 15]. The latter papers also consider the related problem of constructing small "discrepancy sets" for geometric and combinatorial rectangles. We note that a polynomial (in all parameters) "hitting set" for such rectangles was constructed in [247].

t-universal set generators. An ϵ-bias generator, for $\epsilon < 2^{-t}$, yields a t-universal set generator. The latter generator outputs sequences such that in every subsequence of length t all possible 2^t patterns occur (for at least one possible seed). Such generators have many applications (cf., for example, [256, 65]).

3.6.3 Random Walks on Expanders

By **expander graphs** (or **expanders**) of degree d and eigenvalue bound $\lambda < d$, we mean an infinite family of d-regular graphs, $\{G_n\}_{n \in S}$ ($S \subseteq \mathbb{N}$), so that G_n is a d-regular graph over n vertices and the absolute value of all eigenvalues, save the biggest one, of the adjacency matrix of G_n is bounded above by λ.[22] Actually, we are interested in explicit constructions of such graphs, by which we mean that there exists a polynomial-time algorithm that on input n (in binary), a vertex $v \in G_n$ and an index $i \in \{1, ..., d\}$, returns the ith neighbor of v. (We also require that the set S for which G_n's exist is sufficiently "tractable" – say that given any $n \in \mathbb{N}$ one may efficiently find $s \in S$ so that $n \leq s < 2n$.) Many explicit constructions of expanders were

---

[20] Both bounds are derived from the Norm2 bound (of $\epsilon(k)$) on the difference vector (i.e., the difference between the two probability vectors). See [169, Chap. 1].

[21] In the corresponding result for the max-norm distance, it suffices to have $k = O(\log(t(k)/\epsilon(k)) + \log\log \ell(k))$.

[22] This algebraic definition is related to the combinatorial definition of expansion in which one requires that any (not too big) set of vertices in the graph have relatively a large set of strict neighbors (i.e., is "expanding"). See [12] for a lower bound of expansion in terms of $(d-\lambda)/d$ and [8] for the converse. We stress that the back-and-forth translation is not tight, and note that in some applications (alas not those discussed in the current text) the incurred loss when going from the combinatorial definition to the algebraic one is crucial.

given, starting in [258] and culminating in the optimal construction of [249] (where $\lambda = 2\sqrt{d-1}$ and S is somewhat complex[23]). We prefer to use the construction of [159], where $S = \{n^2 : n \in \mathbb{N}\}$, alas it is not optimal.

An important discovery of Ajtai, Komlos, and Szemerédi [4] is that random walks on expander graphs provide a good approximation to repeated independent attempts to hit any arbitrary fixed subset of sufficient density (within the vertex set). The importance of this discovery stems from the fact that a random walk on an expander can be generated using much fewer random coins than required for generating independent samples in the vertex set. That is, generating a random walk of (edge) length ℓ on a d-regular n-vertex graph requires only $\log_2 n + \ell \cdot \log_2 d$ random bits (rather than $(\ell+1) \cdot \log_2 n$ random bits requires to produce independent random samples). Precise formulations of the above discovery were given in [4, 108, 223, 177] culminating in the optimal analysis of [226, Sec. 6].

Theorem 3.23 (Expander Random Walk Theorem [226, Cor. 6.1]): *Let $G = (V, E)$ be an expander graph of degree d and eigenvalue bound λ. Let W be a subset of V and $\rho \overset{\text{def}}{=} |W|/|V|$. Then the fraction of random walks (in G) of (edge) length ℓ which stay within W is at most*

$$\rho \cdot \left(\rho + (1 - \rho) \cdot \frac{\lambda}{d} \right)^\ell$$

Thus, a random walk on an expander is "pseudorandom" with respect to the property of hitting dense sets (i.e., the set $V \setminus W$ above).

Definition 3.24 (the hitting test): *A distribution on sequences over $\{0, 1\}^b$ is (ϵ, δ)-hitting if for any (target) set $T \subseteq \{0, 1\}^b$ of cardinality at least $\epsilon \cdot 2^b$, with probability at least $1 - \delta$, at least one of the elements of a sequence drawn from this distribution hits T.*

Using Theorem 3.23 and the explicit expanders of [159], we have

Proposition 3.25 (The Expander Random Walk Generator):[24] *Let $t \in \mathbb{N}$ be a sufficiently large constant and $d = 2^t$.[25] Let $b, \ell, \ell' : \mathbb{N} \mapsto \mathbb{N}$ so that $b(k)$ is even, $k = b(k) + \ell'(k) \cdot t$, and $\ell(k) = (1 + \ell'(k)) \cdot b(k)$. Let $G_{2^{b(k)}}$ be a d-regular expander graph of [159] and let $\Gamma_\sigma(v)$ denote the vertex reached from v when*

---

[23] Unfortunately, the [249] construction works for a relatively non-trivial set S (i.e., the elements of S are of the form $p \cdot (p^2 - 1)/2$, where p is prime). A relaxation to prime powers is presented in [10, Sec. II].

[24] The common parameterization starts with parameters b and ℓ'. Given a uniformly chosen seed of length $b + O(\ell')$, one can efficiently and deterministically generate a random sequence of $\ell' + 1$ strings, each of length b, which is (ϵ, δ)-hitting for any $\epsilon > 0$ and $\delta = (1 - \frac{\epsilon}{2})^{\ell'+1}$.

[25] t is selected so that the eigenvalue bound of the d-regular expander of [159] is at most $d/2$.

following the edge labeled σ in $G_{2^{b(k)}}$. Then, for $v_0 \in \{0,1\}^{b(k)}$ and σ_i's in $\{0,1\}^t$,

$$G(v_0, \sigma_1, ..., \sigma_{\ell'(k)}) = (v_0, v_1, ..., v_{\ell'(k)})$$

where $v_i = \Gamma_{\sigma_i}(v_{i-1})$, induces a distribution on sequences over $\{0,1\}^{b(k)}$ which is (ϵ, δ)-hitting for any $\epsilon > 0$ and $\delta = (1 - \frac{\epsilon}{2})^{\ell'(k)+1}$.

Expander random-walk generators have been used in a variety of areas (e.g., PCP and the non-approximability of Max-Clique [20, 209][26], and cryptography [177]). In addition, they seem an important tool in the design of various types of "pseudorandom" objects; see below.

3.6.4 Samplers

In this subsection we stretch the pseudorandomness paradigm even further. Except for the case of *averaging samplers* (briefly discussed at the end), the distinguishability test discussed below consists of two components – a fixed algorithm and an arbitrary function, where the former is designed so that no function can distinguish (in a certain sense) the output sequence of the generator from a uniformly selected sequence. In fact, we will combine the above algorithm and the generator into one entity called a sampler. (Another aspect in which samplers deviate from the generators discussed above is in the aim to minimize, rather than maximize, the length of the output sequence. Still, one aims to maximize the block-length, denoted n below.) A reader who is confused by this paragraph is encouraged to forget it, for the time being, and get back to it after reading through the entire subsection.

In many settings repeated sampling is used to estimate the average value of a huge set of values. Namely, there is a value function ν defined over a huge domain, say $\nu : \{0,1\}^n \mapsto [0,1]$, and one wishes to approximate $\bar{\nu} \stackrel{\text{def}}{=} \frac{1}{2^n} \sum_{x \in \{0,1\}^n} \nu(x)$ without having to inspect the value of ν on the entire domain. The obvious thing to do is to sample the domain at random, and obtain such an approximation from the values of ν on the sample points. It turns out that certain "pseudorandom" sequences of sample points may serve almost as well as truly random sequences of sample points.

Formal Setting. It is essential to have the range of ν be bounded (or else no reasonable approximation may be possible). Our convention of having $[0,1]$ be the range of ν is adopted for simplicity, and the problem for other (predetermined) ranges can be treated analogously. Our notion of approximation depends on two parameters: *accuracy* (denoted ϵ) and *error probability* (denoted δ). We wish to have an algorithm which with probability at least $1 - \delta$, gets within ϵ of the correct value. This leads to the following definition.

---

[26] See discussion in [40, Sec. 11.1].

Definition 3.26 (sampler): *A* sampler *is a randomized algorithm that on input parameters* n *(length),* ϵ *(accuracy) and* δ *(error), and oracle access to* ANY *function* $\nu : \{0,1\}^n \mapsto [0,1]$, *outputs, with probability at least* $1 - \delta$, *a value that is at most* ϵ *away from* $\bar{\nu} \stackrel{\text{def}}{=} \frac{1}{2^n} \sum_{x \in \{0,1\}^n} \nu(x)$. *Namely,*

$$\Pr[|\text{sampler}^\nu(n, \epsilon, \delta) - \bar{\nu}| > \epsilon] < \delta$$

where the probability is taken over the internal coin tosses of the sampler.

A non-adaptive sampler *is a sampler which consists of two deterministic algorithms – a* sample generating *algorithm,* G, *and a* evaluation algorithm, V. *On input* n, ϵ, δ *and a random seed, algorithm* G *generates a sequence of queries, denoted* $s_1, ..., s_m \in \{0,1\}^n$. *Algorithm* V *is given the corresponding* ν-*values (i.e.,* $\nu(s_1), ..., \nu(s_m)$) *and outputs an estimate to* $\bar{\nu}$.

We are interested in "the complexity of sampling" quantified as a function of the parameters n, ϵ and δ. Specifically, we will consider three complexity measures: The sample complexity (i.e., the number of oracle queries made by the sampler); the randomness complexity (i.e., the length of the random seed used by the sampler); and the computational complexity (i.e., the running-time of the sampler). We say that a sample is efficient if its running-time is polynomial in the total length of its queries (i.e., polynomial in both its sample complexity and in n). We will focus on efficient samplers. Furthermore, we will focus on efficient samplers which have optimal (up-to a constant factor) sample complexity, and will be interested in having the randomness complexity be as low as possible.

All positive results refer to non-adaptive samplers, whereas the lower bound hold for general samplers. For more details see [171].

The naive sampler. The straightforward method (or the naive sampler) consists of *uniformly and independently* selecting sufficiently many sample points (queries), and outputting the average value of the function on these points. Using Chernoff Bound one easily determines that $O(\frac{\log(1/\delta)}{\epsilon^2})$ sample points suffice. The naive sampler is optimal (up-to a constant factor) in its sample complexity, but is quite wasteful in randomness.

It is known that $\Omega(\frac{\log(1/\delta)}{\epsilon^2})$ samples are needed in any sampler, and that that samplers which make $s(n, \epsilon, \delta)$ queries require randomness at least $n + \log_2(1/\delta) - \log_2 s(n, \epsilon, \delta) - O(1)$ (cf., [85]). These lower bounds are tight (as demonstrated by non-explicit and inefficient algorithms [356]). These facts guide our quest for improvements which is aimed at finding more randomness-efficient ways of *efficiently* generating sample sequences which can be used in conjunction with an appropriate evaluation algorithm V. (We stress that V need not necessarily take the average of the values of the sampled points.)

The pairwise-independent sampler. Here we use the pairwise-independence generator (of Section 3.6.1) to generate sample points, and use the natural evaluation algorithm (which outputs the average of the values of these points). Pairwise-independent sampling yields a great saving in the randomness complexity [103]: Specifically, for constant $\delta > 0$, the Pairwise-Independent Sampler is optimal up-to a constant factor in both its sample and randomness complexities. In general, it uses $2n$ random bits and a sample of size $O(1/\delta\epsilon^2)$. Thus, for small δ (i.e., $\delta = o(1)$), it is wasteful in sample complexity.

The Median-of-Averages sampler. A new idea is required for going further, and a relevant tool – random walks on expander graphs (see above) – is needed too. In [37], the Pairwise-Independent Sampler is combined with the Expander Random Walk Generator to obtain a new sampler. Loosely speaking, the new sampler uses a random walk on an expander to generate a sequence of $t \stackrel{\text{def}}{=} O(\log(1/\delta))$ (related) seeds for t invocations of the Pairwise-Independent Sampler. Each of these invocations returns an ϵ-close approximation with probability at least 0.9. The Expander Random Walk Theorem is used to show that, with probability at least $1 - \exp(-t) = 1 - \delta$, most of these t invocations return an ϵ-close approximation. Thus, the median value is an (ϵ, δ)-approximation to the correct value. The resulting sampler, called the Median-of-Averages Sampler, has sample complexity $O(\frac{\log(1/\delta)}{\epsilon^2})$ and randomness complexity $2n + O(\log(1/\delta))$, which is optimal up-to a constant factor in both complexities.

Further improvements. A sampler which improves over the pairwise-independent sampler is presented in [194]. Maintaining the sample complexity of the latter (i.e., $O(1/\delta\epsilon^2)$), the new sampler has randomness complexity $n + O(\log(1/\delta\epsilon))$ (rather than $2n$). Actually, the general problem of approximating the value of functions mapping to $[0, 1]$ can be efficiently reduced to the problem of estimating the fraction of 1's in Boolean functions (cf., [339]). For the Boolean case, the sampler amounts to picking a random vertex in a suitable expander graph and using the neighbor set as a sample (i.e., one outputs the average over these neighbors). This sampler is identical to a hitting procedure previously suggested in [231], but the analysis is slightly more involved here. Combining this new sampler with the Median-of-Averages idea, one obtains a sampler of sample complexity $O(\frac{\log(1/\delta)}{\epsilon^2})$ and randomness complexity $n + O(\log(1/\delta)) + O(\log(1/\epsilon))$.

Averaging Samplers. Averaging (a.k.a. Oblivious) samplers are non-adaptive samplers in which the evaluation algorithm is the natural one – that is it outputs the average of the values of the sampled points. Interestingly, averaging samplers have applications for which ordinary (non-adaptive) samplers do not suffice (cf., [53, 356, 340]). An averaging sampler of sample complexity $\text{poly}((n/\epsilon) \cdot \log(1/\delta))$ and randomness complexity $(1 + \alpha) \cdot (n + \log_2(1/\delta))$, for every $\alpha > 0$, is presented in [356].

3.6.5 Dispersers, Extractors and Weak Random Sources

In this subsection we stretch the pseudorandomness paradigm even more than in the previous subsection. Specifically, when we consider (below) Weak Random Sources, we will in some sense say that these sources are pseudorandom with respect to one specific algorithm, which is actually designed in purpose so that to be fooled by such sources. Actually, the technical tools defined below (i.e., dispersers and extractors) can also be viewed as pseudorandom generators of a type similar to the generating algorithm of an non-adaptive sampler. (As samplers, these generators are non-standard in the sense that they aim to minimize the length, denoted $\ell'(k)$, of the output sequence. Still, again, the aim is to maximize the block length, denoted $b(k)$.)[27] Our presentation is quite terse; for more details see [284].

Definition 3.27 (disperser):[28] *Let $b, \ell', m : \mathbb{N} \mapsto \mathbb{N}$ and $\epsilon : \mathbb{N} \mapsto [0, 1]$, and $F = \{f_k\}_{k \in \mathbb{N}}$ be a function ensemble with $f_k : \{0, 1\}^k \times \{1, ..., \ell'(k)\} \mapsto \{0, 1\}^{b(k)}$. The ensemble F is called an (m, ϵ)-disperser if for every set $S \subset \{0, 1\}^k$ of cardinality $2^{m(k)}$, the set*

$$\{f_k(s, i) : s \in S \land i \in \{1, ..., \ell'(k)\}\}$$

contains at least $(1 - \epsilon(k)) \cdot 2^{b(k)}$ elements.

A disperser as above generates hitting sequences in the natural way. That is, $G(s) \stackrel{\text{def}}{=} (f_{|s|}(s, 1), ..., f_{|s|}(s, \ell'(|s|)))$, induces a distribution of sequences over $\{0, 1\}^{b(k)}$ which is (ϵ, δ)-hitting for $\delta(k) = 2^{-(k - m(k))}$. Put in other words, for any set $T \subset \{0, 1\}^b$ of cardinality greater than $\epsilon \cdot 2^b$ there exists at most 2^m possible s's such that the set $\{f_{|s|}(s, i) : i \in \{1, ..., \ell'\}\}$ does not intersect T. The following stronger notion, called an extractor, guarantees that there exists at most 2^m possible s's such that the number of i's satisfying $f(s, i) \in T$ does not approximate the density of T. More generally, define the *min-entropy* of a distribution X to be the minimum of $\log_2(1/\Pr[X = x])$, taken over x's in the support of X. Then an extractor is defined so that, for all X's of sufficient min-entropy and for a uniformly selected $i \in \{1, ..., \ell'\}$, the expected value of $f(X, i) \in T$ approximates the density of T. This means that in such cases, the distribution $f(X, i)$ is close to the uniform distribution.

---

[27] Again, we deviate from the standard presentation, where the block length $n \stackrel{\text{def}}{=} b(k)$ is viewed as the principle parameter, and the ultimate goal is to have explicit constructions with $k, \ell'(k) = \text{poly}(n)$ (for $m(k)$ and $\epsilon(k)$ as small as possible). Clearly, $m(k) > n - O(\log n)$, assuming $\epsilon(k)$ is bounded away from 1.

[28] A popular presentation is in terms of (regular) bipartite graphs. The graph corresponding to f_k, will have 2^k vertices each of degree ℓ' on one side, and 2^b vertices on the other side. It will be required that every set of 2^m vertices on the first side will neighbor all but an ϵ fraction of the other side vertices.

Definition 3.28 (extractor):[29] *Let* $b, \ell', m : \mathbb{N} \mapsto \mathbb{N}$, $\epsilon : \mathbb{N} \mapsto [0, 1]$, *and* $F = \{f_k\}_{k \in \mathbb{N}}$ *be as in Definition 3.27. The ensemble F is called an (m, ϵ)-extractor if for every random variable $X \in \{0, 1\}^k$ of min-entropy $m(k)$ and U being uniformly distributed over $\{1, ..., \ell'(k)\}$, the random variable $f(X, U)$ is at most $\epsilon(k)$-away[30] from the uniform distribution over $\{0, 1\}^{b(k)}$.*

An extractor as above yields a (non-adaptive) sampler consisting of the sample generating algorithm $G(s) \overset{\text{def}}{=} (f_{|s|}(s, 1), ..., f_{|s|}(s, \ell'(|s|)))$, and the standard evaluation algorithm which takes the average. This sampler approximates the average of any function up-to ϵ with error probability δ, where $\delta(k) = 2^{-(k-m(k))}$. We comment that every family of Universal$_2$ Hash functions yields an extractor, alas with poor parameters (typically, with $m(k) = k - \Theta(\log \ell'(k))$, whereas below we mention constructions with much smaller value of $m(k)$).

Explicit constructions. By an explicit disperser (resp., extractor) we mean one for which there exists a polynomial-time evaluation algorithm (which on input s and i returns $f_{|s|}(s, i)$). The known results exhibit a trade-off between the various parameters (i.e., the functions b, m, ℓ' and ϵ); see [284]. Here we mention only three results, fixing a function ϵ so that $\epsilon(k) = 1/\text{poly}(k)$. Typically, the goal is to maximize the function b and minimize the functions m and ℓ'.

Theorem 3.29 (explicit dispersers [318]): *For every $\alpha > 0$ there exists $\beta > 0$ so that for $m(k) = \lfloor k^\alpha \rfloor$ and $b(k) = \lfloor k^\beta \rfloor$, explicit (m, ϵ)-dispersers with range $\{0, 1\}^{b(\cdot)}$ and $\ell'(k) = \text{poly}(k)$ exists.*

Theorem 3.30 (explicit extractors [356, 337]): *Explicit (m, ϵ)-extractors with range $\{0, 1\}^{b(\cdot)}$ exist in two cases*

1. *For any $\alpha > 0$, there exists $\beta > 0$ so that $m(k) = \lfloor \alpha k \rfloor$, $b(k) = \lfloor \beta k \rfloor$ and $\ell'(k) = \text{poly}(k)$ (cf., [356]).*
2. *For any integer c, for every $\alpha > 0$ there exists $\beta > 0$ so that $m(k) = \lfloor k^\alpha \rfloor$, $b(k) = \lfloor k^\beta \rfloor$ and $\ell'(k) = \lfloor k^{\log^{(c)} k} \rfloor$, where $\log^{(c)}$ denotes the logarithm function iterated c times (cf., [337]).*

As should be clear from the above discussion, these explicit dispersers (resp., extractors) yield efficient hitting sequence generators (resp., samplers) of very low randomness complexity. Specifically, the error probability of these generators (resp., samplers) is $2^{-(k-m(k))}$, which is extremely close to the "optimum" of 2^{-k}. Turning the table around, these dispersers (resp., extractors)

---

[29] Again, a popular presentation is in terms of (regular) bipartite graphs. Here it will be required, as a special case, that every set of 2^m vertices on the first side will have approximately the same number of edges to all but an ϵ fraction of the other side vertices.

[30] Distance between distributions is defined as their variation distance; that is, the distance between Y_1 and Y_2 is defined as $\frac{1}{2} \sum_y |\Pr[Y_1 = y] - \Pr[Y_2 = y]|$.

can be used to simulate one-sided error (resp., two-sided error) randomized algorithm using very defective (or weak) random sources – see below.

Simulations using Weak Random Sources. Given a randomized algorithm, our goal is to convert it into a robust randomized algorithm which maintain its performance also when its random choices are implemented by a defective (or weak) random source [347]. Such transformations, for increasingly weaker (or more general) types of defective sources have appeared in [347, 104, 355, 318, 337, 14] (omitting quite a few papers). The weakest source considered, hereafter denoted an (k, m)-source, supplies a single k-bit string of min-entropy m. Specifically, using the explicit disperser mentioned above, for every $\alpha > 0$, one can simulate any one-sided error randomized algorithm by an algorithm of polynomial-related complexity which uses any (k, k^α)-source, where k is the length of the random bit sequence required by the new algorithm. The analogous result for two-sided error randomized algorithms [14] was obtained by using an additional idea which is beyond the scope of this chapter.

3.7 Concluding Remarks

In this section we further discuss the computational approach to randomness, provide a historical account of its evolution, and propose some open problems. Figure 3.1 depicts some of the various notions of pseudorandom generators discussed above.

TYPE	distinguisher	generator	stretch; i.e., $\ell(k)$	comments
archetypic.	poly(k)-time	poly(k)-time	poly(k)	OW Assum.[31]
derand. BPP	$2^{k/O(1)}$-time	$2^{O(k)}$-time	$2^{k/O(1)}$	E.C. Assum.[31]
space	$s(k)$-space	$O(k)$-space	$2^{k/O(s(k))}$	runs in time
robust	$k/O(1)$-space	$O(k)$-space	poly(k)	poly$(k) \cdot \ell(k)$
t-wise indep.	"t-wise"	poly$(k) \cdot \ell(k)$-time	$2^{k/O(t)}$	(e.g., pairwise)
small bias	"ϵ-bias"	poly$(k) \cdot \ell(k)$-time	$2^{k/O(1)} \cdot \epsilon(k)$	
expander	hitting	poly$(k) \cdot \ell(k)$-time	$(1 + \ell'(k)) \cdot b(k)$	
rand. walk	$(0.5, 2^{-\ell'(k)/O(1)})$-hitting for $\{0,1\}^{b(k)}$, with $\ell'(k) = (k - b(k))/O(1)$.			

Fig. 3.1. Pseudorandom generators at a glance

---

[31] By OW Assum. we denote the assumption that one-way functions exists; whereas by E.C. Assum. we denote the seemingly weaker assumption by which the class $\mathcal{E}$ does not have subexponential-size circuits (cf., Theorem 3.16).

3.7.1 Discussion

We discuss several conceptual aspects of the above computational approach to randomness.

Behavioristic versus Ontological. The behavioristic nature of the computational approach to randomness is best demonstrated by confronting this approach with the Kolmogorov-Chaitin approach to randomness. Loosely speaking, a string is *Kolmogorov-random* if its length equals the length of the shortest program producing it. This shortest program may be considered the "true explanation" to the phenomenon described by the string. A Kolmogorov-random string is thus a string which does not have a substantially simpler (i.e., shorter) explanation than itself. Considering the simplest explanation of a phenomenon may be viewed as an ontological approach. In contrast, considering the effect of phenomena on certain objects, as underlying the definition of pseudorandomness, is a behavioristic approach. Furthermore, there exist probability distributions which are not uniform (and are not even statistically close to a uniform distribution) that nevertheless are indistinguishable from a uniform distribution (by any efficient method) [351, 180]. Thus, distributions which are ontologically very different, are considered equivalent by the behavioristic point of view taken in the definitions above.

A relativistic view of randomness. Pseudorandomness is defined above in terms of its observer. We have considered several classes of observers, ranging from general efficient (i.e., polynomial-time) observers to very restricted types of observers (e.g., the linear or hitting tests). Each such class gave rise to a different notion of pseudorandomness. Furthermore, the general paradigm explicitly aims at distributions which are not uniform and yet are indistinguishable from such. Thus, our entire approach to pseudorandomness is relativistic and subjective (i.e., depending on the abilities of the observer).

Randomness and Computational Difficulty. Pseudorandomness and computational difficulty play dual roles: The general paradigm of pseudorandomness relies on the fact that putting computational restrictions on the observer gives rise to distributions which are not uniform and still cannot be distinguished from uniform. Furthermore, many of the construction of pseudorandom generators have relied on either conjectures or facts regarding computations which are hard for certain classes. For example, one-way functions were used to construct the archetypical pseudorandom generators (i.e., those working in polynomial-time and fooling all polynomial-time observers), and the fact that PARITY is hard for polynomial-size constant-depth circuits was used to generate sequences which fool such circuits.

3.7.2 Historical Perspective

Our presentation, which views vastly different notions as incarnations of a general paradigm of pseudorandomness, is indeed non-standard (especially, when referred to the special-purpose generators). This unified view appears only in retrospect, and is less evident from the actual historical development of the various notions (although some links can be traced, as done below).

The archetypical pseudorandom generators. The key concept of *computational indistinguishability* was suggested by Goldwasser and Micali in the context of defining secure encryption schemes [198]. The general definition is due to Yao [351], who also proved – using the hybrid technique of [198] – that defining pseudorandom generators as producing sequences which are computationally indistinguishable from uniform is equivalent as defining them to produce unpredictable sequences. The latter definition is due to Blum and Micali who were the first to construct pseudorandom generators based on some simple intractability assumption (in their case the intractability of Discrete Logarithm problem over prime fields) [71]. Their work also introduces basic paradigms which were used in all subsequent improvements (cf., [351, 245, 181, 211]): Basing pseudorandomness on hard problems, the usage of hard-core predicates (defined in [71]), and the iteration paradigm. The fundamental result by which pseudorandom generators exist if and only if one-way functions exist is due to Håstad, Impagliazzo, Levin and Luby [211]. Pseudorandom functions were defined and first constructed by Goldreich, Goldwasser and Micali [174].

Derandomization of time-complexity classes. As observed by Yao [351], a non-uniformly strong notion of pseudorandom generators yields improved derandomization of time-complexity classes. A key observation of Nisan [281, 286] is that whenever a pseudorandom generator is used this way, it suffices to require that the generator runs in time exponential in its seed length, and so the generator may have running-time greater than the distinguisher (representing the algorithm to be derandomized). This observation underlines the construction of Nisan and Wigderson [281, 286], and is the basis for further improvements culminating in [221].

Space Pseudorandom Generators. As stated in the first paper on the subject [4],[32] this research direction was inspired by the de-randomization result obtained via use of archetypical pseudorandom generators. The latter result (necessarily) depends on intractability assumptions, and so the objective was to find classes of algorithms for which derandomization is possible without relying on intractability assumptions. (This objective was achieved before for constant-depth circuits [5].) Fundamentally different constructions of space pseudorandom generators were given in [4, 28, 282, 287], where Nisan's Generator [282] improves over all the previous ones, and the Nisan-Zuckerman Generator [287] is incomparable to Nisan's.

---

[32] This paper is more frequently cited for the Expander Random Walk technique which it has introduced.

Special Purpose Generators. The various generators presented in Section 3.6 were certainly not inspired by the archetypical pseudorandom generator (nor even by a generic notion of pseudorandomness). Since their development is rather technical in nature, we see little point to repeat or elaborate on the credits gives in the text of Section 3.6.

3.7.3 Open Problems

As mentioned above, Theorem 3.13 is currently established via an impractical and complex construction. An alternative construction of (archetypical) pseudorandom generators based on an *arbitrary* one-way function would be most appreciated.

The intractability assumptions used in non-trivial derandomizations of $\mathcal{BPP}$ seem to be getting increasingly weaker (cf., from [351] to [221]). Can one place $\mathcal{BPP}$ in a deterministic class lower than $\mathcal{EXP} = \mathrm{Dtime}(2^{\mathrm{poly}})$, without using any assumptions?

In the area of space-robust pseudorandom generators *the* open problems are well-known – improving over the parameters of the known generators (i.e., of Theorems 3.18 and 3.19), and over the known derandomization results for $\mathcal{RL}$ or for undirected connectivity (cf., [319] and [16], respectively).

With respect to special-purpose generators, two famous open problems are providing constructions, polynomial in all parameters, for small discrepancy sets w.r.t combinatorial rectangles and for extractors (so to improve over [15] and [356, 337], respectively).

We also mention the related (to discrepancy) open problem of providing a deterministic polynomial-time approximation of the number of satisfying assignment of DNF formulae. Recall that randomized polynomial-time and deterministic quasi-polynomial-time (relative error) approximators are known (cf. [229, 281, 253]).

Acknowledgments

Thanks to Amnon Ta-Shma and Luca Trevisan for commenting on earlier versions of this chapter.

A. Background on Randomness and Computation

This appendix contains some basic background on probability theory (Section A.1) and on computational complexity theory (Sections A.2 and A.3). This background is assumed throughout the book. The appendix also contains an intuitive description of the basic settings of Cryptography (Section A.4). Familiarity with these settings is assumed in Chapter 1.

A.1 Probability Theory – Three Inequalities

The following probabilistic inequalities are often used in the analysis of randomized algorithms, and refer to random variables which are assigned real values (e.g., the success probability of a single run of an algorithm). All inequalities refer to random variables which are assigned values within some interval. The most basic inequality, known as *Markov Inequality*, provides bounds on the probability mass which may be assigned to values which are much above (resp., below) the expected value. Specifically,

Markov Inequality: Let X be a *non-negative* random variable and v a positive real number. Then

$$\Pr\left(X \geq v\right) \leq \frac{\mathrm{Exp}(X)}{v}$$

Equivalently, $\Pr(X \geq r \cdot \mathrm{Exp}(X)) \leq \frac{1}{r}$.

Proof:

$$
\begin{aligned}
\mathrm{Exp}(X) &= \sum_x \Pr(X=x) \cdot x \\
&\geq \sum_{x<v} \Pr(X=x) \cdot 0 + \sum_{x \geq v} \Pr(X=x) \cdot v \\
&= \Pr(X \geq v) \cdot v
\end{aligned}
$$

The claim follows. □

Markov inequality is typically used in cases one knows very little about the distribution of the random variable. It suffices to know its expectation and at least one bound on the range of its values. Typical applications are

1. Let X be a random variable so that $\mathrm{Exp}(X) = \mu$ and $X \leq 2\mu$. Then $\Pr[X \leq \frac{\mu}{2}] \leq \frac{2}{3}$.
2. Let $0 < \epsilon, \delta < 1$, and Y be a random variable ranging in the interval $[0, 1]$ such that $\mathrm{Exp}(Y) = \delta + \epsilon$. Then $\Pr[Y \geq \delta + \frac{\epsilon}{2}] > \frac{\epsilon}{2}$.

Using Markov's inequality, one can obtain a "possibly stronger" bound for the deviation of a random variable from its expectation. This bound, called Chebyshev's inequality, is useful provided one has additional knowledge concerning the random variable (specifically a good upper bound on its variance).

Chebyshev's Inequality: Let X be a random variable, and $\delta > 0$. Then

$$\Pr\left[|X - \mathrm{Exp}(X)| \geq \delta\right] \leq \frac{\mathrm{Var}(X)}{\delta^2}$$

Proof: We define a random variable $Y \overset{\text{def}}{=} (X - \mathrm{Exp}(X))^2$, and apply Markov inequality. We get

$$\Pr\left[|X - \mathrm{Exp}(X)| \geq \delta\right] = \Pr\left[(X - \mathrm{Exp}(X))^2 \geq \delta^2\right]$$
$$\leq \frac{\mathrm{Exp}((X - \mathrm{Exp}(X))^2)}{\delta^2}$$

and the claim follows. $\square$

Chebyshev's inequality is particularly useful in the analysis of the error probability of approximation via repeated sampling. It suffices to assume that the samples are picked in a pairwise independent manner.

Corollary (Pairwise Independent Sampling): Let $X_1, X_2, ..., X_n$ be pairwise independent random variables with the identical expectation, denoted μ, and identical variance, denoted σ^2. Then

$$\Pr\left[\left|\frac{\sum_{i=1}^n X_i}{n} - \mu\right| \geq \delta\right] \leq \frac{\sigma^2}{\delta^2 n}$$

The X_i's are said to be *pairwise independent* if for every $i \neq j$ and all a, b, it holds that $\Pr[X_i = a \wedge X_j = b]$ equals $\Pr[X_i = a] \cdot \Pr[X_j = b]$.

Proof: Define the random variables $\overline{X}_i \overset{\text{def}}{=} X_i - \mathrm{Exp}(X_i)$. Note that the $\overline{X}_i$'s are pairwise independent, and each has zero expectation. Applying Chebyshev's inequality to the random variable defined by the sum $\sum_{i=1}^n \frac{X_i}{n}$, and using the linearity of the expectation operator, we get

$$\Pr\left[\left|\sum_{i=1}^n \frac{X_i}{n} - \mu\right| \geq \delta\right] \leq \frac{\mathrm{Var}\left(\sum_{i=1}^n \frac{X_i}{n}\right)}{\delta^2}$$
$$= \frac{\mathrm{Exp}\left(\left(\sum_{i=1}^n \overline{X}_i\right)^2\right)}{\delta^2 \cdot n^2}$$

Now (again using the linearity of Exp)

$$\text{Exp}\left(\left(\sum_{i=1}^{n}\overline{X}_i\right)^2\right) = \sum_{i=1}^{n}\text{Exp}\left(\overline{X}_i^2\right) + \sum_{1\le i\neq j\le n}\text{Exp}\left(\overline{X}_i\overline{X}_j\right)$$

By the pairwise independence of the $\overline{X}_i$'s, we get $\text{Exp}(\overline{X}_i\overline{X}_j) = \text{Exp}(\overline{X}_i) \cdot \text{Exp}(\overline{X}_j)$, and using $\text{Exp}(\overline{X}_i) = 0$, we get

$$\text{Exp}\left(\left(\sum_{i=1}^{n}\overline{X}_i\right)^2\right) = n \cdot \sigma^2$$

The corollary follows.   □

Using pairwise independent sampling, the error probability in the approximation is decreasing linearly with the number of sample points. Using totally independent sampling points, the error probability in the approximation can be shown to decrease exponentially with the number of sample points. (The random variables $X_1, X_2, ..., X_n$ are said to be *totally independent* if for every sequence $a_1, a_2, ..., a_n$ it folds that $\Pr[\wedge_{i=1}^{n}X_i = a_i]$ equals $\prod_{i=1}^{n}\Pr[X_i = a_i]$.)

The bounds quote below are (weakenings of) a special case of the *Martingale Tail Inequality* which suffices for our purposes. The first bound, commonly referred to as *Chernoff Bound*, concerns 0-1 random variables (i.e., random variables which are assigned as values either 0 or 1).

Chernoff Bound: Let $p \le \frac{1}{2}$, and $X_1, X_2, ..., X_n$ be independent 0-1 random variables so that $\Pr[X_i = 1] = p$, for each i. Then for all δ, $0 < \delta \le p(1 - p)$, we have

$$\Pr\left[\left|\frac{\sum_{i=1}^{n}X_i}{n} - p\right| > \delta\right] < 2 \cdot e^{-\frac{\delta^2}{2p(1-p)}\cdot n}$$

We will usually apply the bound with a constant $p \approx \frac{1}{2}$. In this case, n independent samples give an approximation which deviates by ϵ from the expectation with probability δ which is exponentially decreasing with $\epsilon^2 n$. Such an approximation is called an (ϵ, δ)-*approximation*, and can be achieved using $n = O(\epsilon^{-2} \cdot \log(1/\delta))$ sample points. It is important to remember that the sufficient number of sample points is polynomially related to ϵ^{-1} and logarithmically related to δ^{-1}. So using poly(n) many samples the error probability (i.e. δ) can be made exponentially vanishing (as a function in n), but the accuracy of the estimation can be only bounded above by any fixed polynomial fraction. A generalization of Chernoff Bound, which useful in the approximations of the expectation of a general random variable (not necessarily 0-1), is given below.

Hoefding Inequality: Let $X_1, X_2, ..., X_n$ be n independent random variables with identical probability distribution, each ranging over the (real) interval $[a, b]$, and let μ denote the expected value of each of these variables. Then,

$$\Pr\left[\left|\frac{\sum_{i=1}^{n} X_i}{n} - \mu\right| > \delta\right] < 2 \cdot e^{-\frac{2\delta^2}{(b-a)^2} \cdot n}$$

A.2 Computational Models and Complexity Classes

In this section, we briefly recall the definitions of complexity classes such as $\mathcal{P}$, $\mathcal{NP}$, $\mathcal{BPP}$, and non-uniform $\mathcal{P}$ (i.e., $\mathcal{P}/\text{poly}$), and the concept of oracle machines. All these classes are defined in terms of worst-case complexity. We shortly discuss average-case complexity at the end of this section.

A.2.1 P, NP, and More

A conservative approach to computing devices associates efficient computations with the complexity class $\mathcal{P}$. Jumping ahead, we note that the approach taken in this book is a more liberal one in that it allows the computing devices to use coin tosses.

Definition A.1 (P): $\mathcal{P}$ *is the class of languages which can be recognized by* (deterministic) *polynomial-time Turing machines* (algorithms).

Likewise, the complexity class $\mathcal{NP}$ is associated with computational problems having solutions that, once given, can be efficiently tested for validity. It is customary to define $\mathcal{NP}$ as the class of languages which can be recognized by a non-deterministic polynomial-time machine. A more fundamental interpretation of $\mathcal{NP}$ is given by the following equivalent definition.

Definition A.2 (NP): *A language L is in $\mathcal{NP}$, if there exists a Boolean relation $R_L \subseteq \{0,1\}^* \times \{0,1\}^*$ and a polynomial $p(\cdot)$ such that R_L can be recognized in* (deterministic) *polynomial-time and $x \in L$ if and only if there exists a y such that $|y| \leq p(|x|)$ and $(x, y) \in R_L$. Such a y is called a* witness *for membership of $x \in L$.*

Thus, $\mathcal{NP}$ consists of the set of languages for which there exist short proofs of membership that can be efficiently verified. It is widely believed that $\mathcal{P} \neq \mathcal{NP}$, and settling this conjecture is certainly the most intriguing open problem in Theoretical Computer Science. Generalizing the above definitions we have:

Definition A.3 (Dtime and Ntime): *Let $t : \mathbb{N} \mapsto \mathbb{N}$. Then $L \in \text{Dtime}(t)$* (resp., $L \in \text{Ntime}(t)$) *if there exists a deterministic* (resp., *non-deterministic*) *Turing machine for deciding L so that for any input x the machine runs for at most $t(|x|)$ steps.*

Clearly, $\mathcal{P} = \bigcup_c \text{Dtime}(p_c)$ and $\mathcal{NP} = \bigcup_c \text{Ntime}(p_c)$, where $p_c(n) \stackrel{\text{def}}{=} n^c$.

A.2.2 Probabilistic Polynomial-Time

The basic thesis underlying our discussion is the association of "efficient" computations with probabilistic polynomial-time computations. Namely, we will consider as efficient only randomized algorithms (i.e., probabilistic Turing machines) whose running time is bounded by a polynomial in the length of the input. Such algorithms (machines) can be viewed in two equivalent ways.

One way of viewing randomized algorithms is to allow the algorithm to make random moves (i.e., "toss coins"). Formally this can be modeled by a Turing machine in which the transition function maps pairs of the form ($\langle$state$\rangle$, $\langle$symbol$\rangle$) to two possible triples of the form ($\langle$state$\rangle$, $\langle$symbol$\rangle$, $\langle$direction$\rangle$). The next step of such a machine is determined by a random choice of one of these triples. Namely, to make a step, the machine chooses at random (with probability one half for each possibility) either the first triple or the second one, and then acts accordingly. These random choices are called the *internal coin tosses* of the machine. The output of a probabilistic machine, M, on input x is not a string, but rather a random variable assuming strings as possible values. This random variable, denoted $M(x)$, is induced by the internal coin tosses of M. By $\Pr[M(x)=y]$ we mean the probability that machine M on input x outputs y. The probability space is that of all possible outcomes for the internal coin tosses of M, taken with uniform probability distribution. The last sentence is slightly more problematic than it seems. The simple case is when, on input x, machine M always makes the same number of internal coin tosses (independent of their outcome). (In general, the number of coins tossed may depend on the outcome of previous coin tosses.) Still, since we only consider machines of bounded run-time, we may assume (without loss of generality) that the number of coin tosses made by M on input x is independent of their outcome, and is denoted by $t_M(x)$. We denote by $M_r(x)$ the output of M on input x when r is the outcome of its internal coin tosses. Then, $\Pr[M(x)=y]$ is merely the fraction of $r \in \{0,1\}^{t_M(x)}$ for which $M_r(x) = y$. Namely,

$$\Pr\left[M(x)=y\right] = \frac{|\{r \in \{0,1\}^{t_M(x)} : M_r(x)=y\}|}{2^{t_M(x)}}$$

The second way of looking at randomized algorithms is to view the outcome of the internal coin tosses of the machine as an auxiliary input. Namely, we consider deterministic machines with two inputs. The first input plays the role of the "real input" (i.e., x) of the first approach, while the second input plays the role of a possible outcome for a sequence of internal coin tosses. Thus, the notation $M(x,r)$ corresponds to the notation $M_r(x)$ used above. In the second approach one considers the probability distribution of $M(x,r)$, for any *fixed* x and a uniformly chosen $r \in \{0,1\}^{t_M(x)}$. Pictorially, here the coin tosses are not "internal" but rather supplied to the machine by an "external" coin tossing device.

Before continuing, let us remark that one should not confuse the fictitious model of "non-deterministic" machines with the model of probabilistic
machines. The first is an unrealistic model which is useful for talking about
search problems the solutions to which can be efficiently verified (e.g., the
definition of $\mathcal{NP}$), while the second is a realistic model of computation.

In the sequel, unless otherwise stated, a *probabilistic polynomial-time Turing machine* means a probabilistic machine that always (i.e., independently
of the outcome of its internal coin tosses) halts after a polynomial (in the
length of the input) number of steps. It follows that the number of coin
tosses of a probabilistic polynomial-time machine M is bounded by a polynomial, denoted T_M, in its input length. Finally, without loss of generality,
we assume that on input x the machine always makes $T_M(|x|)$ coin tosses.

Thesis: *Efficient computations correspond to computations that can be car-
ried out by probabilistic polynomial-time Turing machines.*

A complexity class capturing these computations is the class, denoted
$\mathcal{BPP}$, of languages recognizable (with high probability) by probabilistic
polynomial-time machines. The probability refers to the event "the machine
makes correct verdict on string x".

Definition A.4 (Bounded-Probability Polynomial-time — $\mathcal{BPP}$): $\mathcal{BPP}$ *is
the class of languages which can be recognized by a probabilistic polynomial-
time machine (i.e., randomized algorithm). We say that L is recognized by
the probabilistic polynomial-time machine M if*

- *For every $x \in L$ it holds that $\Pr[M(x)=1] \geq \frac{2}{3}$.*
- *For every $x \notin L$ it holds that $\Pr[M(x)=0] \geq \frac{2}{3}$.*

The phrase "bounded-probability" indicates that the success probability is
bounded away from $\frac{1}{2}$. In fact, substituting in Definition A.4 the constant $\frac{2}{3}$
by any other constant greater than $\frac{1}{2}$ does not change the class defined. More
generally, $L \in \mathcal{BPP}$ if there exists a polynomial-time computable (threshold) function $t : \mathbb{N} \mapsto [0,1]$, a positive polynomial $p(\cdot)$ and a probabilistic
polynomial-time machine, M, such that

- For every $x \in L$ it holds that $\Pr[M(x)=1] > t(|x|) + \frac{1}{p(|x|)}$.
- For every $x \notin L$ it holds that $\Pr[M(x)=1] < t(|x|) - \frac{1}{p(|x|)}$.

(The fact that such L is in $\mathcal{BPP}$ can be proven using Chebyshev's Inequality.)
On the other hand, using Chernoff's Bound one can prove that, for every $L \in
\mathcal{BPP}$ and every positive polynomial p, there exists a probabilistic polynomial-
time machine, M, such that

- For every $x \in L$ it holds that $\Pr[M(x)=1] \geq 1 - 2^{-p(|x|)}$
- For every $x \notin L$ it holds that $\Pr[M(x)=0] \geq 1 - 2^{-p(|x|)}$

The class $\mathcal{BPP}$ captures two-sided error probabilistic polynomial-time computations. Two analogous classes which capture (complementary) one-sided error computations are $\mathcal{RP}$ and co$\mathcal{RP}$.

Definition A.5 (RP and coRP):

- $L \in \mathcal{RP}$ *if there exists a probabilistic polynomial-time machine M so that*
  - *For every $x \in L$ it holds that $\Pr[M(x)=1] \geq \frac{1}{2}$.*
  - *For every $x \notin L$ it holds that $\Pr[M(x)=0] = 1$.*
- $L \in$ co$\mathcal{RP}$ *if there exists a probabilistic polynomial-time machine M so that*
  - *For every $x \in L$ it holds that $\Pr[M(x)=1] = 1$.*
  - *For every $x \notin L$ it holds that $\Pr[M(x)=0] \geq \frac{1}{2}$.*

Analogously to the case of $\mathcal{BPP}$, these classes remain robust when substituting the constant $\frac{1}{2}$ by either $\frac{1}{p(|x|)}$ or $1-2^{-p(|x|)}$, for every positive polynomial p. Clearly, $\mathcal{P} \subseteq \mathcal{RP} \subseteq \mathcal{BPP}$ and $\mathcal{RP} \subseteq \mathcal{NP}$.

A.2.3 Non-Uniform Polynomial-Time

A stronger model of efficient computation is that of non-uniform polynomial-time. This model will be used only in the negative way; namely, for saying that even such machines cannot do something.

A *non-uniform polynomial-time "machine"* is a pair $(M, \bar{a})$, where M is a two-input polynomial-time machine and $\bar{a} = a_1, a_2, ...$ is an infinite sequence of strings such that $|a_n| = \text{poly}(n)$. For every x, we consider the computation of machine M on the input pair $(x, a_{|x|})$. Intuitively, a_n may be thought as an extra "advice" supplied from the "outside" (together with the input $x \in \{0,1\}^n$). We stress that machine M gets the same advice (i.e., a_n) on all inputs of the same length (i.e., n). Intuitively, the advice a_n may be useful in some cases (i.e., for some computations on inputs of length n), but it is unlikely to encode enough information to be useful for all 2^n possible inputs.

Another way of looking at non-uniform polynomial-time "machines" is to consider an infinite sequence of machines, $M_1, M_2, ...$ so that both the length of the description of M_n and its running time on inputs of length n are bounded by polynomial in n (fixed for the entire sequence). Machine M_n is used only on inputs of length n. Note the correspondence between the two ways of looking at non-uniform polynomial-time. The pair $(M, (a_1, a_2, ...))$ of the first definition gives rise to an infinite sequence of machines $M_{a_1}, M_{a_2}, ...,$ where $M_{a_{|x|}}(x) \stackrel{\text{def}}{=} M(x, a_{|x|})$. On the other hand, a sequence $M_1, M_2, ...$ (as in the second definition) gives rise to the pair $(U, (\langle M_1 \rangle, \langle M_2 \rangle, ...))$, where U is a universal Turing machine and $\langle M_n \rangle$ is the description of machine M_n (i.e., $U(x, \langle M_{|x|} \rangle) = M_{|x|}(x)$).

In the first sentence of the current subsection, non-uniform polynomial-time has been referred to as a stronger model than probabilistic polynomial-time. This statement is valid in many contexts (e.g., language recognition as in Theorem 1 below). In particular it will be valid in all contexts we discuss in this book. So we have the following informal "meta-theorem"

Meta-Theorem: Whatever can be achieved by probabilistic polynomial-time machines can be achieved by non-uniform polynomial-time "machines".

The Meta-Theorem is clearly wrong if one thinks of the task of tossing coins... So the meta-theorem should not be understood literally. It is merely an indication of real theorems that can be proven in reasonable cases. Let's consider the context of language recognition.

Definition A.6 (P/poly): *The complexity class non-uniform polynomial-time, denoted $\mathcal{P}/\text{poly}$, is the class of languages L which can be recognized by "non-uniform polynomial-time machines". Namely, $L \in \mathcal{P}/\text{poly}$ if there exists an infinite sequence of machines $M_1, M_2, \ldots$ satisfying*

1. *There exists a polynomial $p(\cdot)$ such that, for every n, the description of machine M_n has length bounded above by $p(n)$.*
2. *There exists a polynomial $q(\cdot)$ such that, for every n, the running time of machine M_n on each input of length n is bounded above by $q(n)$.*
3. *For every n and every $x \in \{0,1\}^n$, machine M_n accepts x if and only if $x \in L$.*

Note that the non-uniformity is implicit in the lack of a requirement concerning the construction of the machines in the sequence. It is only required that these machines exist. In contrast, if one augments Definition A.6 by requiring the existence of a polynomial-time algorithm that on input 1^n (n presented in unary) outputs the description of M_n then one gets a cumbersome way of defining $\mathcal{P}$. On the other hand, it is obvious that $\mathcal{P} \subseteq \mathcal{P}/\text{poly}$ (in fact strict containment can be proven by considering non-recursive unary languages). Furthermore,

Theorem A.7 $\mathcal{BPP} \subseteq \mathcal{P}/\text{poly}$.

Proof: Let M be a probabilistic machine recognizing $L \in \mathcal{BPP}$. Let $\xi_L(x) \overset{\text{def}}{=} 1$ if $x \in L$ and $\xi_L(x) = 0$ otherwise. Then, for every $x \in \{0,1\}^*$,

$$\Pr[M(x) = \xi_L(x)] \geq \frac{2}{3}$$

Assume, without loss of generality, that on each input of length n, machine M uses the same number, $m = \text{poly}(n)$, of coin tosses. Let $x \in \{0,1\}^n$. Clearly, we can find for each $x \in \{0,1\}^n$ a sequence of coin tosses $r \in \{0,1\}^m$ such that $M_r(x) = \xi_L(x)$ (in fact most sequences r have this property). But can one sequence $r \in \{0,1\}^m$ fit all $x \in \{0,1\}^n$? Probably not (provide an example!). Nevertheless, we can find a sequence $r \in \{0,1\}^n$ which fits $\frac{2}{3}$ of all the x's of length n. This is done by a counting argument (which asserts that if $\frac{2}{3}$ of the r's are good for each x then there is an r which is good for at least $\frac{2}{3}$ of the x's). However, this does not give us an r which is good for all $x \in \{0,1\}^n$. To get such an r we have to apply the above argument on a machine M' with exponentially vanishing error probability. Such a machine is guaranteed by the alternative formulation of $\mathcal{BPP}$ (given above). Namely, for every $x \in \{0,1\}^*$,

$$\Pr[M'(x) = \xi_L(x)] > 1 - 2^{-|x|}$$

Applying the argument now we conclude that there exists an $r \in \{0,1\}^m$, denoted r_n, which is good for *more than* a $1 - 2^{-n}$ fraction of the $x \in \{0,1\}^n$. It follows that r_n is good for all the 2^n inputs of length n. Machine M' (viewed as a deterministic two-input machine) together with the infinite sequence $r_1, r_2, \ldots$ "constructed" as above, demonstrates that L is in $\mathcal{P}/\text{poly}$. $\square$

Finally, let us mention a more convenient (and standard) way of viewing non-uniform polynomial-time. This is via (non-uniform) families of polynomial-size Boolean circuits. A *Boolean circuit* is a directed acyclic graph with internal nodes marked by elements in $\{\wedge, \vee, \neg\}$. Nodes with no in-going edges are called *input nodes*, and nodes with no outgoing edges are called *output nodes*. A node marked $\neg$ may have only one in-going edge. Computation in the circuit begins by placing input bits on the input nodes (one bit per node) and proceeds as follows. If the children of a node (of in-degree d) marked $\wedge$ have values $v_1, v_2, \ldots, v_d$ then the node gets the value $\bigwedge_{i=1}^{d} v_i$. Similarly for nodes marked $\vee$ and $\neg$. The output of the circuit is read from its output nodes. The *size* of a circuit is the number of its edges. A *polynomial-size circuit family* is an infinite sequence of Boolean circuits, $C_1, C_2, \ldots$ such that, for every n, the circuit C_n has n input nodes and size $p(n)$, where $p(\cdot)$ is a polynomial (fixed for the entire family). Clearly, the computation of a Turing machine M on inputs of length n can be simulated by a single circuit (with n input nodes) having size $O((|\langle M \rangle| + n + t(n))^2)$, where $t(n)$ is a bound on the running time of M on inputs of length n. Thus, a non-uniform sequence of polynomial-time machines can be simulated by a non-uniform family of polynomial-size circuits. The converse is also true, since machines with polynomial description length can incorporate polynomial-size circuits and simulate their computations in polynomial-time. The thing which is nice about the circuit formulation is that there is no need to repeat the polynomiality requirement twice (once for size and once for time) as in the two formulations above.

A.2.4 Oracle Machines

The original utility of oracle machines in complexity theory is to capture notions of reducibility (see below). In the context of cryptography oracle machines are used for a seemingly different purpose – to model an adversary which may use a cryptosystem in course of its attempt to break it. A third usage of oracle machines is in the context of defining pseudorandom functions (cf., Section 3.3.4). Lastly, oracle machines are used when defining Probabilistically Checkable Proof (PCP) systems (cf., Section 2.4).

Definition A.8 (oracle machines): *A* (deterministic/probabilistic) oracle machine *is a* (deterministic/probabilistic) *Turing machine with an additional tape, called the* oracle tape, *and two special states, called* oracle invocation *and* oracle appeared. *The* computation of the deterministic oracle machine *M on input x and access to the oracle* $f : \{0,1\}^* \mapsto \{0,1\}^*$, *denoted* $M^f(x)$, *is defined by the* successive configuration relation. *For configurations with state different from "oracle invocation" the next configuration is defined as usual. Let γ be a configuration in which the state is "oracle invocation" and the contents of the oracle tape is q. Then the configuration following γ is identical to γ, except that the state is "oracle appeared" and the contents of the oracle tape is $f(q)$. The string q is called M's query and $f(q)$ is called the* oracle reply. *The computation of a probabilistic oracle machine is defined analogously.*

We stress that the running time of an oracle machine is the number of steps made during its computation, and that the oracle's reply on each query is obtained in a single step. As stated above, oracle machines are used to define general notions of reducibility. Below, a language L is identified with its characteristic function χ_L, where $\chi_L(x) = 1$ if $x \in L$ and $\chi_L(x) = 0$ otherwise.

Definition A.9 (Turing or Cook reducibility): *A language L_1 is said to be* reducible *to a language L_2 if there exists a probabilistic polynomial-time oracle machine M so that*

– *For every $x \in L_1$ it holds that $\Pr[M^{L_2}(x)=1] \geq \frac{2}{3}$.*
– *For every $x \notin L_1$ it holds that $\Pr[M^{L_2}(x)=0] \geq \frac{2}{3}$.*

In both cases L_2 is viewed as a Boolean function so that $L_2(q) = 1$ iff $x \in L_2$.

Unless stated differently, whenever we say a reduction, we mean a Turing reduction as above. A more restricted notion of a reduction follows.

Definition A.10 (many-to-one or Karp reducibility): *A language L_1 is said to be* many-to-one reducible *to a language L_2 if there exists a polynomial-time compatible function, f, so that for every x*

$$x \in L_1 \text{ if and only if } f(x) \in L_2$$

A.2.5 Space Bounded Machines

In contrast to all the above, we now consider complexity classes defined by the space consumed by the computation, rather than by its time. The space complexity of algorithms (Turing machines) is defined as the space consumed *by the computation itself*; that is, not counting the input and output. Thus, one considers Turing machines with one-way (read-only) input-tape, one-way (write-only) output-tape, and several auxiliary work tapes. The space complexity of a computation is defined as the number of cells scanned on the work-tapes.

For logarithmic (and higher) space complexity,[1] the space complexity remains invariant if we add the logarithm of the input length. This convention allows to equate the space complexity of a computation with the logarithm of the number of possible configurations of the computation on a specific input (where each configuration consists of the contents of the work-tapes and the locations of all heads on all tapes).[2]

The most popular space-complexity classes are $\mathcal{L}$ and $\mathcal{NL}$ – the set of all languages recognizable by deterministic, resp., non-deterministic, machines of logarithmic space complexity. Clearly, $\mathcal{L} \subseteq \mathcal{NL} \subseteq \mathcal{P}$. The definition of the analogous randomized classes, denoted $\mathcal{BPL}$ and $\mathcal{RL}$, is more problematic (cf., Section 3.5).

A.2.6 Average-Case Complexity

In contrast to all the above, we now discuss *average-case* (rather than worst-case) complexity. Thus the domain consists of *distributional problems* which are pairs consisting of a traditional computational problem (e.g., a decision problem) coupled with a probability distribution. The issues at hand are which distributions to allow (since placing no restrictions on the distributions may collapse average-case complexity to its worst-case analogue) and how to define efficient computation (since, surprisingly, the naive definitions suffer from fundamental problems). A theory of average-case complexity, addressing these issues, has been initiated by Levin [244] (cf., [172]).

Much of the material in this book presupposes not only that $\mathcal{P} \neq \mathcal{NP}$ but also the ability to efficiently generate hard-on-the-average instances of a computational problem. Furthermore, the latter assumption is even strengthened by requiring that one may efficiently generate instance-solution pairs, so that the instances are hard (on the average) to solve. To be specific, let us consider the problem of finding NP-witnesses. That is, let L, p and R_L be as in Definition A.2. We will assume that for *some* probabilistic polynomial-time algorithm G, with $G(1^n)$ ranging over $R_L \cap (\{0,1\}^n \times \{0,1\}^{p(n)})$, the following holds: For *any* probabilistic polynomial-time algorithm A, any positive polynomial q, and all sufficiently large n's

$$\Pr[(X_n, A(X_n)) \in R_L] < \frac{1}{q(n)}$$

---

[1] Here we consider only logarithmic and higher space complexity.

[2] For a fixed input, there is no need to include the contents of the input-tape as it remains invariant throughout the computation.

where X_n is distributed as the first element of $G(1^n)$. The above assumption is equivalent to assuming the existence of one-way functions (as defined in Sections 1.2.1 and 3.3.3).[3]

A.3 Complexity Classes – Glossary

$\mathcal{AC}_0$ – The set of languages recognized by constant-depth, polynomial-size circuits.

AM – Typically, this denotes the class of languages having an interactive proof system in which the verifier sends a single uniformly chosen message. This class equals $\mathcal{IP}(2)$.

BPL – In analogy to $\mathcal{BPP}$, this is the class of languages recognized by *probabilistic polynomial-time two-sided error* machines of logarithmic space-complexity.

BPP – See Definition A.4.

coAM – $L \in \text{co}\mathcal{AM}$ if $\overline{L} \in \mathcal{AM}$.

coNP – $L \in \text{co}\mathcal{NP}$ if $\overline{L} \in \mathcal{NP}$.

coRP – See Definition A.5. ($L \in \text{co}\mathcal{RP}$ if $\overline{L} \in \mathcal{RP}$.)

CZK – The set of languages having a *computational* zero-knowledge proof system. Also denoted $\mathcal{ZK}$.

Dtime – See Definition A.3.

E – A shorthand for $\bigcup_c \text{Dtime}(e_c)$, where $e_c(n) \overset{\text{def}}{=} 2^{cn}$.

EXP – A shorthand for $\bigcup_c \text{Dtime}(e_c)$, where $e_c(n) \overset{\text{def}}{=} 2^{n^c}$.

IP and $\mathcal{IP}(\cdot)$ – See Definition 2.2.

L – The set of languages recognizable by deterministic machines of logarithmic space-complexity.

MA – The class of languages having an interactive proof system in which the verifier sends no messages, and merely uses randomization for its decision regarding the prover's message. This class equals $\mathcal{IP}(1)$.

Ntime – See Definition A.3.

NEXP – A shorthand for $\bigcup_c \text{Ntime}(e_c)$, where $e_c(n) \overset{\text{def}}{=} 2^{n^c}$.

NL – The set of languages recognizable by *non-deterministic* machines of logarithmic space-complexity.

NP – See Definition A.2.

NP-complete – L is NP-complete if it is both NP-hard and in $\mathcal{NP}$.

NP-hard – In a narrow sense, L is NP-hard if any language in $\mathcal{NP}$ is Karp reducible to it. In a wide sense, a computational problem (not necessarily a language recognition problem), is NP-hard if any language in $\mathcal{NP}$ is (Turing) reducible to it.

---

[3] Given G as above we define a one-way function by mapping the coins used by $G(1^n)$ to the first element of its output. Conversely, given a one-way function f, we define the NP-relation $R = \{(f(w), w) : w \in \{0,1\}^*\}$ and consider the generator G that on input 1^n uniformly selects $w \in \{0,1\}^n$ and outputs $(f(w), w)$.

P – See Definition A.1.

P/poly – See Definition A.6.

PCP and $\mathcal{PCP}(\cdot, \cdot)$ – See Definition 2.10.

PH – The polynomial-time hierarchy, defined as $\bigcup_{c \in \mathbb{N}} \Sigma_c^P$, where $\Sigma_{c+1}^P = \mathcal{NP}^{\Sigma_c^P}$ and $\Sigma_1^P = \mathcal{NP}$. For a class $\mathcal{C}$, the class $\mathcal{NP}^{\mathcal{C}}$ contains languages for which there exists a non-deterministic polynomial-time oracle machine M so that, given oracle access to some language in $\mathcal{C}$, there exists an accepting computation of M on input x if and only if x is in the language. See [94].

PSPACE – The set of languages recognizable by (deterministic) machines of polynomial space-complexity.

PZK – See Definition 2.5.

RL – In analogy to $\mathcal{RP}$, this is the class of languages recognized by *probabilistic polynomial-time one-sided error* machines of logarithmic space-complexity.

RP – See Definition A.5.

SC – The set of languages recognizable by *polynomial-time* deterministic machines of *polylogarithmic space-complexity*.

SZK – The set of languages having a *statistical* zero-knowledge proof system. This relaxes *perfect* zero-knowledge as defined in Definition 2.5.

ZK – The set of languages having a (*computational*) zero-knowledge proof system. Also denoted $\mathcal{CZK}$. Unless $\mathcal{PSPACE} = \mathcal{AM}$ (which is most unlikely), $\mathcal{ZK}$ is a strict generalization of *statistical* zero-knowledge. See Section 2.3.

A.4 Some Basic Cryptographic Settings

In this section we briefly review four basic problems of cryptography, giving rise to the notions of private-key and public-key encryption and signatures. We also review the definition of the RSA and Rabin functions, which we view as prime candidates for (trapdoor) one-way functions.

The purpose of this section is merely to familiarize the reader with some basic notions, and so the presentation is quite informal. Actual definitions of one-way functions, secure encryption schemes and unforgeable signatures are provided in Chapter 1.

A.4.1 Encryption Schemes

The problem of providing *secret communication over insecure media* is the traditional and most basic problem of cryptography. The setting of this problem consists of two parties communicating through a channel which is possibly tapped by an adversary. The parties wish to exchange information with each other, but keep the "wiretapper" as ignorant as possible regarding the contents of this information. Loosely speaking, an encryption scheme is a protocol allowing these parties to communicate *secretly* with each other. Typically, the encryption scheme consists of a pair of algorithms. One algorithm, called *encryption*, is applied by the sender (i.e., the party sending a message), while the other algorithm, called *decryption*, is applied by the receiver. Hence, in order to send a message, the sender first applies the encryption algorithm to the message, and sends the result, called the *ciphertext*, over the channel. Upon receiving a ciphertext, the other party (i.e., the receiver) applies the decryption algorithm to it, and retrieves the original message (called the *plaintext*).

In order for the above scheme to provide secret communication, the communicating parties (at least the receiver) must know something which is not known to the wiretapper. (Otherwise, the wiretapper can decrypt the ciphertext exactly as done by the receiver.) This extra knowledge may take the form of the decryption algorithm itself, or some parameters and/or auxiliary inputs used by the decryption algorithm. We call this extra knowledge the *decryption key*. Note that, without loss of generality, we may assume that the decryption algorithm is known to the wiretapper, and that the decryption algorithm operates on two inputs – a ciphertext and a decryption key. We stress that the existence of a secret key, not known to the wiretapper, is merely a necessary condition for secret communication.

Evaluating the "security" of an encryption scheme is a very tricky business. A preliminary task is to understand what is "security" (i.e., to properly define what is meant by this intuitive term). Two approaches to defining security are known. The first ("classic") approach is *information theoretic*. It is concerned with the "information" about the plaintext which is "present" in the ciphertext. Loosely speaking, if the ciphertext contains information about the plaintext then the encryption scheme is considered insecure. It has been shown that such high (i.e., "perfect") level of security can be achieved only if the key in use is at least as long as the *total* length of the messages sent via the encryption scheme. The fact, that the key has to be longer than the information exchanged using it, is indeed a drastic limitation on the applicability of such encryption schemes.

The second ("modern") approach, followed in the current book, is based on *computational complexity*. This approach is based on the observation that it does not matter *whether the ciphertext contains information about the plaintext*, but rather *whether this information can be* efficiently extracted. In other words, instead of asking whether it is *possible* for the wiretapper to extract specific information, we ask whether it is *feasible* for the wiretapper to extract this information. It turns out that the new (i.e., "computational complexity") approach offers security even if the key is much shorter than the total length of the messages sent via the encryption scheme.

The computational complexity approach allows the introduction of concepts and primitives which cannot exist under the information theoretic approach. A typical example is the concept of *public-key encryption schemes*. Note that in the above discussion we concentrated on the decryption algorithm and its key. It can be shown that the encryption algorithm must get, in addition to the message, an auxiliary input which depends on the decryption key. This auxiliary input is called the *encryption key*. Traditional encryption schemes, and in particular all the encryption schemes used in the millenniums until the 1980's, operate with an encryption key equal to the decryption key. Hence, the wiretapper in this schemes must be ignorant of the encryption key, and consequently the *key distribution* problem arises (i.e., how can two parties wishing to communicate over an insecure channel agree on a secret encryption/decryption key). (The traditional solution is to exchange the key through an alternative channel which is secure, though "more expensive to use", for example by a convoy.) The computational complexity approach allows the introduction of encryption schemes in which the encryption key may be given to the wiretapper without compromising the security of the scheme. Clearly, the decryption key in such schemes is different and furthermore infeasible to compute from the encryption key. Such encryption scheme, called *public-key*, have the advantage of trivially resolving the key distribution problem since the encryption key can be publicized.

In contrast, traditional encryption scheme in which the encryption-key equals the description-key are called *private-key* schemes, as in these schemes the encryption-key must be kept secret (rather than be public as in public-key encryption schemes). We note that a full specification of either schemes requires the specification of the way keys are generated; that is, a key-generation (randomized) algorithm which given a security parameter produces a (random) pair of corresponding encryption/decryption keys (which are identical in case of private-key schemes).

A.4.2 Digital Signatures and Message Authentication

The need to discuss "digital signatures" has arise with the introduction of computer communication in business environment (in which parties need to commit themselves to proposals and/or declarations they make). Discussions of "unforgeable signatures" did take place also in previous centuries, but the objects of discussion were handwritten signatures (and not digital ones), and the discussion was not perceived as related to "cryptography". Loosely speaking, a *scheme for unforgeable signatures* requires that

- each user can *efficiently produce his own signature* on documents of his choice;
- every user can *efficiently verify* whether a given string is a signature of another (specific) user on a specific document; but
- *nobody can efficiently produce signatures of other users* to documents they did not sign.

We note that the formulation of unforgeable digital signatures provides also a clear statement of the essential ingredients of handwritten signatures. The ingredients are each person's ability to sign for himself, a universally agreed verification procedure, and the belief (or assertion) that it is infeasible (or at least hard) to forge signatures in a manner that pass the verification procedure. It is hard to assess to what extent do handwritten signatures meet these requirements. In contrast, our discussion of digital signatures will supply precise statements concerning the extend by which digital signatures meet the above requirements. Furthermore, unforgeable digital signature schemes can be constructed based on some reasonable computational assumptions. Loosely speaking, a signature scheme consists of three algorithms corresponding to the key-generation, signing and verification tasks. As in case of encryption, the signing-key is the (secret) information which distincts the legitimate signer from all other users. Analogously to the case of public-key encryption, other users only have the corresponding verification-key allowing them to verify signatures (but not to produce them).

Message authentication

Message authentication is a task related to the setting considered for encryption schemes; that is – communication over an insecure channel. This time, we consider an active adversary which is monitoring the channel and may alter the messages sent on it. The parties communicating through this insecure channel wish to authenticate the messages they send so that their counterpart can tell an original message (sent by the sender) from a modified one (i.e., modified by the adversary). Loosely speaking, a *scheme for message authentication* requires that

— each of the communicating parties can *efficiently produce an authentication tag* to any message of his choice;
— each of the communicating parties can *efficiently verify* whether a given string is an authentication tag of a given message; but
— *no external adversary* (i.e., a party other than the communicating parties) *can efficiently produce authentication tags* to messages not sent by the communicating parties.

Note that in contrast to the specification of signature schemes we do not require universal verification. That is, only the receiver is required to be able to verify the authentication tags, and the fact that the receiver can also produce such tags is of no concern. Thus, schemes for message authentication can be viewed as a private-key version of signature schemes. The difference between the two is that in the setting of message authentication the ability to verify tags may be linked to the ability to authenticate messages, whereas in the setting of signature schemes these abilities are separated (i.e., everybody can verify signatures but only the holder of the signing-key can produce valid signatures). Hence, digital signatures provide a solution to the message authentication problem, but message authentication schemes do not necessarily constitute a digital signature scheme.

A.4.3 The RSA and Rabin Functions

In contrast to some common presentations, we view the RSA and Rabin functions as tools (e.g., candidate one-way functions) rather than as full-fledged utilities (e.g., candidate encryption schemes). Loosely speaking, one-way functions are functions which are easy to compute but hard to invert. Both the RSA and Rabin functions have "trapdoor information", which when given allows to efficiently invert them. (This does not contradict the hardness of inverting postulated above, as it refers to inversion when not given this trapdoor information.)

The conjectured hardness properties of the RSA and Rabin functions are based on the assumption that the Integer Factorization Problem is intractable. In particular, both functions utilize composite numbers which are the product of two large primes and are based on the assumption that it is infeasible to factor such composites. Both the RSA and Rabin functions are actually collections of functions: Each such function is associated with a composite, denoted N, which is the product of two primes, denoted P and Q. Typically, one assumes that $|\log_2 P - \log_2 Q| \le 1$.

The RSA function

A generic function in the RSA collection is determined by a pair, (N, e), where $N = P \cdot Q$ and e is an integer smaller than N and relatively prime to $\phi(N) \stackrel{\text{def}}{=} (P-1) \cdot (Q-1)$. Such a function, denoted $RSA_{N,e}$, is defined over the domain $\{1, ..., N\}$ so that $RSA_{N,e}(x) \stackrel{\text{def}}{=} x^e \bmod N$. Using the fact that e is relatively prime to $\phi(N)$, it can be shown that the function is in fact a permutation over its domain. Furthermore, knowledge of the inverse of e modulo $\phi(N)$, allows to efficiently invert $RSA_{N,e}$. That is, on input (N, d) and y, where $ed \equiv 1 \pmod{\phi(N)}$ and $y = x^e \bmod N$, one can efficiently retrieve x by computing $y^d \bmod N$ (since $(x^e)^d \equiv x^{ed} \equiv x \pmod{N}$, for all x's).

It is widely believed that given (N, e) (but neither d not the factorization of N), it is infeasible to invert $RSA_{N,e}$. Hence, it is conjectured that the RSA collection is a collection of trapdoor (one-way) permutations. However, it is not known whether factoring N can be reduced to inverting $RSA_{N,e}$ (in fact this is a well-known open problem).

The Rabin function

The Rabin collection of functions is defined analogously to the RSA collection, except that the function is squaring modulo N (instead of raising to the power

$e \bmod N$). Namely, $Rabin_N(x) \stackrel{\text{def}}{=} x^2 \bmod N$. This function, however, does not induces a permutation on the multiplicative group modulo N, but is rather a 4-to-1 mapping on this group.

It can be shown that extracting square roots modulo N is computationally equivalent to factoring N (i.e., the two tasks are reducible to one another via probabilistic polynomial-time reductions) [303]. Thus, assuming that factoring is intractable, it is infeasible given N and $y = Rabin_N(x)$ to find a preimage of y. On the other hand, given the factorization of N, it is feasible to find all 4 preimages of y under $Rabin_N$. Hence, assuming that factoring is intractable, the above yields a collection of trapdoor one-way functions.

For a special subclass of the integers, known as *Blum Integers*, the function $Rabin_N(\cdot)$ defined above induces a permutation on the quadratic residues modulo N. We say that r is a *quadratic residue mod N* if there exists an integer x such that $r \equiv x^2 \bmod N$. We denote by Q_N the set of quadratic residues in the multiplicative group mod N, and say that N is a Blum Integer if it is the product of two primes each congruent to 3 mod 4. It can be shown that when N is a Blum integer, each element in Q_N has a unique square root which is also in Q_N, and it follows that in this case the function $Rabin_N(\cdot)$ induces a permutation over Q_N. Hence, assuming that factoring Blum Integers is intractable, the above is a collection of trapdoor (one-way) permutations.

B. Randomized Computations

The purpose of this appendix is to demonstrate the usage of randomization in a variety of computational settings. Our choice is governed by the desire to focus on the randomization aspect of the solution and avoid any complicated details which are due to other aspects of the computational problem. Thus, we avoid any example which requires substantial problem-specific background. We stress that our presentation is merely aimed at *demonstrating* the usage of randomization, and that no attempt was made to present a coherent theory of randomized computation.

Our examples are grouped in three (subjective) categories:

1. Traditional algorithmic problems. Here we consider *randomized algorithms* for graph theoretic problems such as finding a perfect matching, algebraic problems such as testing polynomial identity, and approximation problems such as approximating the number of satisfying assignments to a DNF formula.
2. Traditional complexity questions. Here we present results such as the *randomized reductions* of Approximate Counting to $\mathcal{NP}$, and of SAT to unique-SAT.
3. Distributed and Parallel Computing. Here we consider *randomized procedures* for *distributed tasks* such as Testing String Equality, Byzantine Agreement, and routing in networks.

For a more systematic and much wider exposition, the reader is referred to the textbook *Randomized Algorithms* by Motwani and Raghavan [269].

B.1 Randomized Algorithms

Conspicuous omissions in this category include some of the most well-known randomized algorithms (e.g., many in the domain of computational number theory), as well as the Markov Chain approach to approximate counting. As stated above, the reason for these omissions is that these algorithms either require specialized (and unrelated to randomness) background or are quite involved to present and/or analyze.

B.1.1 Approx. Counting of DNF Satisfying Assignments or, a Twist on Naive Sampling

The problem considered here is to approximate the number of satisfying assignment to a DNF formula up-to a *constant factor*. We note that given ϵ and oracle access to any function $f : \{0,1\}^n \mapsto \{0,1\}$, it is easy to approximate the fraction $|\{x : f(x) = 1\}|/2^n$ up-to an ϵ *additive deviation*. Specifically, a sample of $O(\epsilon^{-2} \log(1/\delta))$ points has average value which, with probability at least $1 - \delta$, is at most ϵ-away from the correct value. However, our aim is to provide relative (rather than absolute) approximation of this fraction (i.e., given $\epsilon > 0$ the task is to approximate the above fraction up-to a $1 \pm \epsilon$ factor).

Let $\varphi = \bigvee_{i=1}^{m} C_i$, where $C_i : \{0,1\}^n \to \{0,1\}$ is a conjunction, be a DNF formula. Actually, we will deal with the more general problem in which we are given (implicitly) m subsets $S_1, ..., S_m \subseteq \{0,1\}^n$ and wish to approximate $|\bigcup_i S_i|$. In our case S_i will be the set of assignments satisfying the conjunction C_i. We make several computational assumptions regarding these sets (letting efficient mean implementable in time polynomial in $n \cdot m$):

1. Given i and x, one can efficiently determine whether $x \in S_i$.
2. Given i, one can efficiently determine $|S_i|$.
3. Given i, one can efficiently generate a uniformly distributed element of S_i.

These assumptions are clearly satisfied in the case $S_i = C_i^{-1}(1)$ considered above. The key observation is that

$$\left| \bigcup_{i=1}^{m} S_i \right| = \sum_{i=1}^{m} \left| S_i \setminus \bigcup_{j<i} S_j \right| \tag{B.1}$$

$$= \sum_{i=1}^{m} |S_i| \cdot \Pr_{s \in S_i} \left[s \notin \bigcup_{j<i} S_j \right] \tag{B.2}$$

and that the probabilities in Eq. (B.2) can be approximated up-to ϵ' (with overwhelming success probability) by taking $\text{poly}(n/\epsilon')$ many samples. This leads to the following algorithm

Algorithm: On input parameters ϵ and δ, set $\epsilon' = \epsilon/m$ and $\delta' = \delta/m$. For $i = 1$ to m do

1. Let $p_i \stackrel{\text{def}}{=} \Pr_{s \in S_i}[s \notin \bigcup_{j<i} S_j]$.

   Using a sample of size $t \stackrel{\text{def}}{=} O((1/\epsilon')^2 \log(1/\delta'))$, approximate p_i by $\widetilde{p}_i$ so that $\Pr[|\widetilde{p}_i - p_i| > \epsilon'] < \delta'$. That is, we uniformly select t samples in S_i, and test for each whether it resides in $\bigcup_{j<i} S_j$.

2. Compute $|S_i|$, and let $a_i \stackrel{\text{def}}{=} \widetilde{p}_i \cdot |S_i|$.

Output the sum of the a_i's.

Analysis: Let $N_i = p_i \cdot |S_i|$. We are interested in the quality of the approximation to $\sum_i N_i$ provided by $\sum_i a_i$. With probability at least $1 - m \cdot \delta'$, we have $a_i = (p_i \pm \epsilon') \cdot |S_i| = N_i \pm \epsilon' \cdot |S_i|$, for all i's, and so $\sum_i a_i = \sum_i N_i \pm \epsilon' \cdot \sum_i |S_i|$. However, $\max_i(|S_i|) \leq |\bigcup_i S_i| = \sum_i N_i$, and so

$$
\sum_{i=1}^{m} a_i = \sum_{i=1}^{m} N_i \pm m \cdot \epsilon' \cdot \max_{1 \leq i \leq m} |S_i|
$$

$$
= (1 \pm m\epsilon') \cdot \sum_{i=1}^{m} N_i = (1 \pm \epsilon) \cdot \sum_{i=1}^{m} N_i
$$

Note that the above approach does not require exact computation of $|S_i|$, nor exact uniform selection in S_i. Instead, ability to approximate $|S_i|$ up-to a factor of $1 \pm \epsilon'$ within time related to $\text{poly}(n/\epsilon')$ suffices. Likewise, it suffice to generate in time related to $\text{poly}(n/\epsilon')$ a distribution which is at most ϵ'-away from the uniform distribution over S_i.

The algorithm presented above is actually a deterministic reduction of the task of approximating the size of one set (in the relative sense) to the task of providing absolute approximations to some fractions. It utilizes the hypothesis that the first set can be expressed as a union of feasibly many sets for which certain natural operations (e.g., deciding membership, approximating the size) can be performed efficiently. Thus, this approach may be applicable to some sets, but not to their complement – which concurs with the general phenomena by which relative approximation may be possible for one quantity, but not for its complement (e.g., it is NP-Hard to approximate the number of UNsatisfying assignment to a DNF formula up-to any factor).

B.1.2 Finding a Perfect Matching
or, on the Loneliness of the Extremum

The problem considered here is to find a perfect matching in a graph. The specific goal is to obtain a fast parallel algorithm, which is the reason we do not follow the standard combinatorial approach of iteratively augmenting the current matching using alternating paths. Instead, we rely on the following Isolation Lemma which asserts that when assigning each edge a random weight, taken from a sufficiently large domain, there is a unique perfect matching of minimum (resp., maximum) weight. The lemma extends to arbitrary set systems.

Lemma B.1 (The Isolation Lemma): *Let* $S_1, S_2, \ldots, S_t \subseteq [m] \stackrel{\text{def}}{=} \{1, 2, \ldots, m\}$ *be distinct sets, and let* $w_1, w_2, \ldots, w_m$ *be independently and uniformly chosen in* $[2m]$. *Then, with probability at least* $1/2$, *there exists a unique* j *so that* $\sum_{i \in S_j} w_i$ *equals* $\min_{k \in [t]}(\sum_{i \in S_k} w_i)$.

In our application $[m]$ corresponds to the set of edges, and the S_i's to perfect matchings in the graph.

Proof: For $i = 1, ..., m$, consider the event E_i defined as the existence of two sets (i.e., S_j's) with minimum weight so that one set contains i and the other set does not contain i. It suffices to show that the probability that E_i occurs is at most $1/2m$. The latter is proven by considering a random process in which the weight of i (i.e., w_i) is selected last.

Suppose that the values of all other w_j's (with $j \neq i$) have already been determined. Let S^- be a set of minimum weight among all sets not containing i, and w^- be its weight (i.e., $w^- \stackrel{\text{def}}{=} \min_{j:i\notin S_j}(\sum_{k\in S_j} w_k)$). Similarly, let S^+ be a set of minimum weight among all sets obtained by omitting i from sets which contain it, and w^+ be its weight (i.e., $w^+ \stackrel{\text{def}}{=} \min_{j:i\in S_j}(\sum_{k\in S_j\setminus\{i\}} w_k)$). Then, event E_i occurs if and only if $w^- = w^+ + w_i$, which happens with probability $1/2m$ if $(w^- - w^+) \in [2m]$, and with probability 0 otherwise. $\square$

Algorithm: On input a bipartite graph $G = (U, V, E)$,

1. For each edge $e \in E$, uniformly and independently select a weight $w_e \in [2m]$, where $m \stackrel{\text{def}}{=} |E|$.
2. Try to compute the value of the minimum weight perfect matching. This is done by computing the determinant of the matrix, denoted A, obtained by setting the (u, v)-entry to 2^{w_e} if $e = (u, v)$ and to 0 if $(u, v) \notin E$. In case the determinant is 0, halt stating that the graph has no perfect matching. Otherwise, the value of the minimum weight perfect matching is set to be the largest i so that the value of the determinant is divisible by 2^i. (The determinant can be computed by a fast parallel algorithm.)
3. For each $e \in E$, try to compute the value of the minimum weight perfect matching among those not containing the edge e. This is done (as above) by computing the determinant of the matrix, denoted A_e, obtained from A by resetting the e-entry to 0. All these computations can be conducted in parallel.
4. A candidate perfect matching is retrieved by including all edges e for which the value (of the min-weight perfect matching) found in Step 3 is different than the one found in Step 2.

The algorithm for general graphs is a variation of the above (and is not described here). Steps 1 and 2 (by themselves) provide a randomized algorithm for determining whether a bipartite graph has a perfect matching.

Analysis: We may assume that the graph has a perfect matching, or else the determinant computed in Step 2 is 0. Assume that the weights (i.e., w_e's) are such that there exists a unique perfect matching of minimum weight. Denote this matching by M and its weight by W. In such a case, the determinant of A is of the form $2^W + r \cdot 2^{W+1}$, where r is an integer (possibly zero). This is so since the determinant sums (possibly with minus sign) the 2-powers of the weights of all perfect matchings, and none can cancel the contribution of the unique minimum weight perfect matching M. Likewise, for every edge e

not in M, the determinant of A_e is of the form $2^W + r \cdot 2^{W+1}$, where again r is an integer. On the other hand, for every edge e in M, the determinant of A_e is either zero or $r \cdot 2^{W+1}$, with r being a non-zero integer.

Comment: It is tempting to think that when selecting weights as above, the minimum weight perfect matching may be uniformly distributed among all perfect matchings.[1] To see that this is not always the case consider a graph in which the set of perfect consists of two types of matchings. There are 2^n matchings of the first type, a generic one having the form $\{e_{2i-\sigma_i} : i = 1, ..., n\}$, where $\sigma_1, ..., \sigma_n \in \{0, 1\}$. There is a single matching of the second type, denoted $\{e_{2n+i} : i = 1, ..., n\}$. We claim that the probability that the minimum weight perfect matching is a specific matching of the first type is exponentially smaller than the probability that the minimum weight perfect matching is the matching of the second type.

This claim holds for weights distributed as above, as well as for several other distributions (e.g., the Normal Distribution). For sake of simplicity, we consider weights uniformly distributed in the interval $[0, 1]$. We first show that, with overwhelmingly high probability, the value of the minimum weight matching among all 2^n matchings of the first type is *at least* cn, where c is any constant smaller than $1/3$ (e.g., $c = 0.32$). This follows by observing that

$$\min_{\sigma_1, ..., \sigma_n \in \{0,1\}} \left(\sum_{i=1}^{n} w_{2i-\sigma_i} \right) = \sum_{i=1}^{n} \min(w_{2i-1}, w_{2i})$$

and that the expected value of each $\min(w_{2i-1}, w_{2i})$ equals $1/3$. On the other hand, the probability that any specific perfect matching (and in particular one of the second type) has weight *less than*, say, $0.31 \cdot n$ is greater than $\frac{0.6^n}{2} = \exp(\Omega(n)) \cdot 2^{-n}$, and so with essentially the same probability the second type matching is of minimum weight among all $2^n + 1$ perfect matchings. This follows by observing that

$$\Pr\left[\sum_{i=1}^{n} w_i < 0.31 \cdot n \right]$$

$$> \quad \Pr[\forall i \, (w_i \leq 0.6)] \cdot \left(1 - \Pr\left[\sum_{i=1}^{n} w_i \geq 0.31 \cdot n \,\middle|\, \forall i \, (w_i \leq 0.6) \right] \right)$$

$$> \quad 0.6^n \cdot \frac{1}{2}$$

where the last inequality uses $\mathrm{Exp}[w_i \,|\, w_i \leq 0.6] = 0.3$.

---

[1] The following text is based on discussions with Madhu Sudan (during March 1998).

B.1.3 Testing Whether Polynomials Are Identical or, on the Discrete Charm of Polynomials

The problem considered here is to determine whether two multi-variant polynomials are identical. We assume that one is given an oracle for the evaluation of each of the polynomials. We further assume that the polynomials are defined over a sufficiently large finite field, denoted F. Finally, let n denote the number of variables in these polynomials.

Algorithm: Given n and black-box access to $p, q : F^n \mapsto F$, uniformly select $r_1, ..., r_n \in F$, and accept if and only if $p(r_1, ..., r_n) = q(r_1, ..., r_n)$.

Analysis: Clearly, if $p \equiv q$ then the algorithm always accepts. The following lemma implies that if p and q are different polynomials, each of total degree at most d, then the algorithm accepts with probability at most $d/|F|$.

Lemma B.2 Let $p : F^n \mapsto F$ be a non-zero polynomial of total degree d. Then

$$\Pr_{r_1,...,r_n}[p(r_1, ..., r_n) = 0] \leq \frac{d}{|F|}$$

Proof: The lemma is proven by induction on n. The base case of $n = 1$ follows immediately by the Fundamental Theorem of Algebra (i.e., the number of distinct roots of a degree d univariant polynomial is at most d). In the induction step, we write p as a polynomial in its first variable. That is,

$$p(x_1, x_2, ..., x_n) = \sum_{i=0}^{d} p_i(x_2, ..., x_n) \cdot x_1^i$$

where p_i is a polynomial of total degree at most $d - i$. Let t be the biggest integer i for which p_i is not identically zero. (We dismiss the case $t = 0$.) Then, using the induction hypothesis, we have

$$\Pr_{r_1,r_2,...,r_n}[p(r_1, r_2, ..., r_n) = 0]$$
$$\leq \Pr_{r_2,...,r_n}[p_t(r_2, ..., r_n) = 0]$$
$$\quad + \Pr_{r_1,r_2,...,r_n}[p(r_1, r_2, ..., r_n) = 0 \,|\, p_t(r_2, ..., r_n) \neq 0]$$
$$\leq \frac{d-t}{|F|} + \frac{t}{|F|}$$

where the second term is bounded by fixing any sequence $r_2, ..., r_n$ for which $p_t(r_2, ..., r_n) \neq 0$ and considering the univariant polynomial $p'(x) \stackrel{\text{def}}{=} p(x, r_2, ..., r_n)$ (which by hypothesis is a non-zero polynomial of degree t). $\square$

Comment: The lesson is that whenever the situation is such that *almost* any choice will do – taking a random choice yields an algorithm with a rigorous performance guarantee. In a sense any randomized algorithm is based on this paradigm, except that here the space of choices seems more straightforward than in any other case. That is, most randomized algorithms are based on introducing a sample space which is not obvious from the problem at hand; whereas here the sample space is the obvious one.

B.1.4 Randomized Rounding Applied to MaxSAT or, on Being Fractionally Pregnant

We slightly deviate from the above style by considering a general methodology. The methodology consists of two steps. First, one presents a linear programming relaxation of an integer program (corresponding to a combinatorial problem). Next, one derives from a solution to the linear program a solution to the integer program, by using the former to determine a probability distribution over solutions to the latter, and picking a solution according to this distribution. We exemplify this methodology by applying it to Max-SAT. Specifically, we consider the task of approximating the maximum number of clauses which can be simultaneously satisfied in a given CNF formula.

Let $\varphi = \bigwedge_{j=1}^{m} C_j$ be a CNF formula, where $C_j = \left(\bigvee_{i \in S_j^+} x_i\right) \vee \left(\bigvee_{i \in S_j^-} \neg x_i\right)$ with $S_j^+, S_j^- \subseteq [n] \stackrel{\text{def}}{=} \{1, ..., n\}$. Abusing notation, we may express Max-SAT as an integer optimization problem in which the task is to maximize $\sum_{j=1}^{m} y_j$ subject to

$$x_i, y_j \in \{0, 1\} \qquad (\forall i, j) \tag{B.3}$$

$$\sum_{i \in S_j^+} x_i + \sum_{i \in S_j^-} (1 - x_i) \geq y_j \qquad (\forall j) \tag{B.4}$$

In the Linear Programming (LP) relaxation one replaces Eq. (B.3) by

$$0 \leq x_i, y_j \leq 1 \quad (\forall i, j) \tag{B.5}$$

Clearly, the value of the LP is lower bounded by the value of the integer program. Given an (optimal) solution, $\hat{x}_i, \hat{y}_j$, to the LP, we randomly derive a solution to the original integer formulation. It will be shown that the expected value of the integer solution is at least $1 - e^{-1}$ times the value of the LP (and hence at least a $1 - e^{-1}$ fraction of the optimum of the integer problem). Specifically, we set $x_i = 1$ with probability $\hat{x}_i$ (and $x_i = 0$ otherwise).

Analysis: Suppose that clause C_j has c_j literals. Then, we will show that the probability that C_j is satisfied by the above randomized rounding (of the above LP solution) is at least

$$\left(1 - \left(1 - \frac{1}{c_j}\right)^{c_j}\right) \cdot \hat{y}_j \geq \left(1 - e^{-1}\right) \cdot \hat{y}_j$$

and so the expected number of satisfied clauses is at least $(1 - e^{-1}) \cdot \sum_j \hat{y}_j$ (as stated above). The above is proven by noting that the probability of the complementary event (i.e., C_j is not satisfied) is

$$\left(\prod_{i \in S_j^+} (1 - \hat{x}_i) \right) \cdot \left(\prod_{i \in S_j^-} \hat{x}_i \right) \tag{B.6}$$

where, by Eq. (B.4), $\sum_{i \in S_j^+} (1 - \hat{x}_i) + \sum_{i \in S_j^-} \hat{x}_i \leq (c_j - \hat{y}_j)$. Eq. (B.6) is maximized when $1 - \hat{x}_i = (c_j - \hat{y}_j)/c_j$ for all $i \in S_j^+$, and $\hat{x}_i = (c_j - \hat{y}_j)/c_j$ for all $i \in S_j^-$. Thus, Eq. (B.6) is bounded above by $\left(1 - \frac{\hat{y}_j}{c_j}\right)^{c_j}$, and the above claim follows.

Comments: Combining the above algorithm with the naive algorithm which uniformly selects a truth assignment, one derives a randomized algorithm of a 3/4-approximation factor. The key observation is that the performance of the LP-based algorithm improves as the clause sizes decrease, whereas the performance of the naive algorithm improves when the sizes increase. In a different vein, we mention that the randomized rounding paradigm has been extended also to *semidefinite* (rather than linear programming) relaxations of combinatorial problems. In fact, improved approximation ratios for various versions of MaxSAT were obtained that way (cf., [164, 228]).

B.1.5 Primality Testing
or, on Hiding Information from an Algorithm

The problem considered here is to decide whether a given number is a prime. The only Number Theoretic facts which we use are:

1. For every prime $p > 2$, each quadratic residue mod p has exactly two square roots mod p (and they sum-up to p).
2. For every (odd and non-integer-power) composite number N, each quadratic residue mod N has at least four square roots mod N.

Our algorithm uses as a black-box an algorithm, denoted R, which given a prime p and a quadratic residue mod p, returns the smallest among the two square roots. There is no guarantee as to what is the output in case the input is not of the above form (and in particular in case p is not a prime).

Algorithm: On input a natural number $N > 2$ do

1. If N is either even or an integer-power then reject.
2. Uniformly select $r \in \{1, ..., N - 1\}$, and set $s \leftarrow r^2 \bmod N$.
3. Let $r' \leftarrow R(N, s)$. If $r' \equiv \pm r \pmod{N}$ then accept else reject.

Analysis: By Fact 1, on input a prime number N, the above algorithm always accepts (since in this case $R(N, r^2 \bmod N) = \pm r$ for any $r \in \{1, ..., N - 1\}$). On the other hand, suppose that N is an odd composite which is not an integer-power. Then, by Fact 2, each quadratic residue s has at least four square roots, and each is equally likely to be chosen at Step 2 (as s yields no information on the specific r). Thus, for every such s, the probability that $\pm R(N, s)$ has been chosen in Step 2 is at most $2/4$. It follows that, on input a composite number, the algorithm rejects with probability at least $1/2$.

Comment: The above analysis presupposes that the algorithm R is always correct when fed with a pair (p, s), where p is prime and s a quadratic residue mod p. In case R has error probability $\epsilon < 1/2$, our algorithm still distinguishes primes from composites (since on the former it accepts with probability at least $1 - \epsilon > 1/2$). We note that efficient randomized algorithms for extracting square roots modulo a prime are known (cf., [29, 269]). Thus, the above establishes that primality can be decided in probabilistic polynomial-time (alas, with two-sided error).

B.1.6 Testing Graph Connectivity via a Random Walk or, the Accidental Tourist Sees It All

The problem considered here is to decide whether a given graph is connected. The aim is to devise an algorithm which does so while using little space (i.e., essentially, as little as needed for storing the identity of a single vertex). This task can be reduced to testing connectivity between any given pair of vertices. Thus, we focus on the task of determining whether two given vertices are connected in a given graph.

Algorithm: On input a graph $G = (V, E)$ and two vertices, s and t, we take a *random walk* of length $O(|V| \cdot |E|)$, starting at vertex s, and test at each step whether vertex t is encountered. By a random walk we mean that, at each step, we uniformly select one of the edges incident at the current vertex and traverse this edge to the other endpoint.

Analysis: We will show that if s is connected to t in the graph G then, with probability at least $1/2$, vertex t is encountered in a random walk starting at s. In the following, we consider the connected component of vertex s, denoted $G' = (V', E')$. For any two vertices, v and u (in V'), we let $T_{u,v}$ be a random variable representing the number of steps taken in a random walk starting at u until v is first encountered. It is easy to see that $E[T_{u,v}] \leq 2|E'|$. Also, letting cover(G') be the expected number of steps in a random walk starting at s and ending when the last of the vertices of V' is encountered, and C be any directed cycle which visits all vertices in G', we have

$$\text{cover}(G') \leq \sum_{(u,v) \in C} E[T_{u,v}]$$
$$\leq |C| \cdot 2|E'|$$

Letting C be a traversal of some spanning tree of G', we conclude that cover$(G') < 4 \cdot |E'| \cdot |V'|$. Thus, with probability at least $1/2$, a random walk of length $8 \cdot |E'| \cdot |V'|$ starting at s visits all vertices of G'.

B.1.7 Finding Minimum Cuts in Graphs
or, Random is Better than Arbitrary

Many algorithms are typically presented in a non-fully specified manner, allowing some choices to be made arbitrarily (in which case these choices are typically made in a way most convenient for implementation). In some cases, replacing the arbitrary choice by a random one yields improved performance. A demonstration of this phenomena follows. The problem considered here is to find the minimum cut in a graph. The randomized algorithm which follows is simpler than the traditional flow-based algorithms, and lends itself to parallel implementation (omitted here).

Algorithm: On input a graph $G = (V, E)$, with $n = |V|$, the algorithm makes $n-2$ random edge *contraction* steps: In each step one selects *uniformly* an edge of the current multi-graph and contracts the two endpoints into one vertex, allowing parallel edges but dropping self-loops which may be created. That is, if (u, v) is the contracted edge of the current graph G' then we replace vertices u and v by a new vertex x, and replace edges of the form (w, v) (resp., (w, u)), where $w \notin \{u, v\}$, by a similar number of edges (w, x). When these $n - 2$ contraction steps are completed, we are left with a multi-graph on two vertices, and just output the number of parallel edges.

Analysis: Suppose that G has a minimum cut $C \subset E$. Then, the probability that no edge of C is contracted in the first step is $\frac{|E|-|C|}{|E|} \geq 1 - \frac{2}{n}$ (since the cut cannot be bigger than the average degree $|C| \leq 2|E|/n$). The question is what happens in subsequent steps. A key observation is that $|C|$ is a lower bound on the average degree of any multi-graph obtained from G by any sequence of edge contractions. Thus, the probability that the $(n - 2)$-step contraction process leaves C intact is at least

$$\prod_{i=1}^{n-2} \left(1 - \frac{2}{n - (i - 1)}\right) = \prod_{i=1}^{n-2} \frac{n - 1 - i}{n + 1 - i} = \frac{2}{n \cdot (n - 1)}$$

Thus, repeating the above algorithm for a quadratic number of times we obtain the minimum cut, with probability at least, say, $2/3$.

Comment: Observe that if the random choices in the above algorithm are replaced by arbitrary choices then the output gives little indication towards the minimum cut in G.

B.2 Randomness in Complexity Theory

In this section we demonstrate the power of randomized reductions (rather than randomized algorithms discussed in the previous section).

B.2.1 Reducing (Approximate) Counting to Deciding or, the Random Sieve

We consider the class $\#\mathcal{P}$ of functions which count the number of NP-witnesses (w.r.t an NP-relation). That is, $f \in \#\mathcal{P}$ if for some NP-relation, R, it holds that $f(x) = |\{y : (x, y) \in R\}|$, for all $x \in \{0, 1\}^*$. We will show that such f can be approximated in probabilistic polynomial-time given oracle to an NP-complete set. The (randomized Cook) reduction uses any efficient family of Universal$_2$ Hash functions[2], as well as the following lemma.

Lemma B.3 (Leftover Hash Lemma [329, 55, 217]):[3] *Let $H_{m,k}$ be a family of Universal$_2$ Hash functions mapping $\{0, 1\}^m$ to $\{0, 1\}^k$, and let $\epsilon > 0$. Let $S \subseteq \{0, 1\}^m$ be arbitrary provided that $|S| \geq \epsilon^{-3} \cdot 2^k$. Then, for all but at most an ϵ fraction of the h's in $H_{m,k}$, it holds that*

$$|\{e \in S : h(e) = 0^k\}| = (1 \pm \epsilon) \cdot \frac{|S|}{2^k}$$

Proof: For a uniformly selected $h \in H_{m,k}$, the random variables $\{h(e)\}_{e \in S}$ are pairwise independent and uniformly distributed over $\{0, 1\}^k$. On top of these $h(e)$'s, we define 0-1 random variables, denoted ζ_e's, so that $\zeta_e = 1$ if $h(e) = 0^k$. Then $\mathrm{Exp}[\zeta_e] = 2^{-k}$ and we need to show that the sum $\sum_{e \in S} \zeta_e$ is concentrated around $|S|/2^k$. Using Chebyshev's Inequality and the fact that the ζ_e's are pairwise independent, we get

$$\Pr\left[\left|\sum_{e \in S} \zeta_e - \frac{|S|}{2^k}\right| > \frac{\epsilon \cdot |S|}{2^k}\right] < \frac{\mathrm{Var}[\sum_{e \in S} \zeta_e]}{(\epsilon |S|/2^k)^2}$$

$$< \frac{|S|/2^k}{\epsilon^2 \cdot (|S|/2^k)^2} \leq \epsilon$$

(Pairwise independence is used in deriving $\mathrm{Var}[\sum_{e \in S} \zeta_e] = \sum_{e \in S} \mathrm{Var}[\zeta_e] < |S| \cdot 2^{-k}$.) □

---

[2] A family of functions mapping $\{0, 1\}^m$ to $\{0, 1\}^k$ is called Universal$_2$ if for a uniformly selected h in the family, the random variables $\{h(e)\}_{e \in \{0,1\}^m}$ are pairwise independent and uniformly distributed over $\{0, 1\}^k$. An efficient family is required to have algorithms for selecting and evaluating functions. A popular example is the family of all linear transformations from $\{0, 1\}^m$ to $\{0, 1\}^k$.

[3] A stronger statement of the lemma, supported by essentially the same proof, refers to an arbitrary random variable X over $\{0, 1\}^m$ satisfying $\Pr[X = x] \leq \epsilon^3 \cdot 2^{-k}$, for every x. The lemma was discovered independently in [55, 217], yet it is an extension of the ideas underlying [329]. The lemma's name was coined in [223].

Reduction: On input $x \in \{0,1\}^n$, the probabilistic polynomial-time oracle machine (for approximating f) sets m to be the length of NP-witness w.r.t the guaranteed R. For every $k = 0, 1, ..., m + 2$ it performs the following experiment n times.

1. Uniformly select $h \in H_{m,k}$, and construct (via Cook's reduction) a CNF formula φ so that φ is satisfiable if and only if there exists a string $y \in \{0,1\}^m$ so that $(x, y) \in R$ and $h(y) = 0^k$.
2. Query the oracle whether φ is satisfiable.

Finally, the machine outputs the smallest non-negative integer k (possibly zero) so that the oracle has answered NO at least $n/2$ times.

Analysis: We analyze the performance of the above machine when it is given oracle access to SAT. Clearly, if $S_x \overset{\text{def}}{=} \{y : (x, y) \in R\}$ has cardinality N then the probability that the machine outputs a number $k \geq L \overset{\text{def}}{=} \lceil \log_2(4N) \rceil$ is exponentially vanishing (since the probability that a uniformly selected $h \in H_{m,L}$ maps some element of S_x to 0^L is at most $1/4$, and so in each iteration with value of $k \geq L$, with probability at least $3/4$, the oracle says NO). On the other hand, using the above lemma, if $N \overset{\text{def}}{=} |S_x| \geq 2^{k+2}$ then for a uniformly selected $h \in H_{m,k}$ with probability at least $3/4$ there exists $y \in S_x$ so that $h(y) = 0^k$. Thus, with overwhelmingly high probability, the output of the oracle machine is at least $\log_2(N/4)$. We conclude that approximating f up-to a factor of 4 is reducible in probabilistic polynomial-time to $\mathcal{NP}$. Higher accuracy – that is, approximation factor of $1 + \frac{1}{p(n)}$, for any fixed positive polynomial p – can be obtained by considering the "direct product function" $F_p(x) \overset{\text{def}}{=} (f(x))^{p(|x|)}$ which counts the number of NP-witnesses w.r.t the NP-relation R_p defined by

$$R_p \overset{\text{def}}{=} \{(x, y_1, ..., y_{p(|x|)}) : \forall i \ (x, y_i) \in R\}$$

A related reduction may be used to reduced SAT (or even "approximating #$\mathcal{P}$") to unique-SAT. By the latter, we mean the promise problem in which the YES-instances are CNF formula having a unique satisfying assignment, and the NO-instances are CNF formula having no satisfying assignment. All that is needed is to notice that in the above reduction, for $k = (\log_2 N) \pm 2$, the reduction produces CNF formula which are typically (i.e., w.p. at least $3/4$) either not satisfiable or have few (say up-to 8) satisfying assignments. Thus, we augment Step 1 as follows. Having produced φ, as above, we produce 8 new formulae, $\psi_1, ..., \psi_8$, so that ψ_i asserts that φ has at least i different satisfying assignments (i.e., $\psi_i(y_1, ..., y_i) = \bigwedge_j \varphi(y_j) \wedge \bigwedge_{1 \leq j < j' \leq i}(y_j < y_{j'}))$. We refer each of these ψ_i to the oracle and use YES as answer if the oracle has answered YES on any of the ψ_i (as this may happen only if φ is indeed satisfiable). Thus, whenever φ has few satisfying assignments, YES will be returned.

B.2.2 Two-sided Error Versus One-sided Error

We consider the extension of the classes $\mathcal{RP}$ and $\mathcal{BPP}$ to promise problems and show that $\mathcal{BPP} = \mathcal{RP}^{\mathcal{RP}}$ (in the extended sense). It is evident that $\mathcal{RP}^{\mathcal{RP}} \subseteq \mathcal{BPP}^{\mathcal{BPP}} = \mathcal{BPP}$ (where the last equality utilizes standard "error reduction"). So we focus on the other direction, considering a BPP-problem with a characteristic function χ (which may be only partially defined over $\{0,1\}^*$) so that for some NP-relation, R, a polynomial p, and for every x on which χ is defined

$$|\{y \in \{0,1\}^{p(|x|)} : R(x,y) \neq \chi(x)\}| < \frac{2^{p(|x|)}}{3p(|x|)}$$

(where $R(x,y) = 1$ if $(x,y) \in R$ and $R(x,y) = 0$ otherwise). We show a randomized one-sided error (Karp) reduction of χ to (the promise problem extension of) co$\mathcal{RP}$.

Reduction: On input $x \in \{0,1\}^n$, the randomized polynomial-time mapping uniformly selects $s_1, ..., s_m \in \{0,1\}^m$, and outputs the pair $(x, \bar{s})$, where $m = p(|x|)$ and $\bar{s} = (s_1, ..., s_m)$.

We define the following co$\mathcal{RP}$ promise problem, denoted Π. The YES-instances, denoted Π_{yes}, are pairs $(x, \bar{s})$ so that for every $r \in \{0,1\}^m$ there exists an i so that $R(x, r \oplus s_i) = 1$. The NO-instances, denoted Π_{no}, are pairs $(x, \bar{s})$ so that for at least half of the possible $r \in \{0,1\}^m$, it holds that $R(x, r \oplus s_i) = 0$ for every i. Clearly, Π is indeed a co$\mathcal{RP}$ promise problem (via an algorithm which uniformly selects r, and computes $R(x, r \oplus s_i)$ for all i's).

Analysis: We claim that the above randomized mapping reduces χ to Π. Suppose first that $\chi(x) = 0$. Then, for every possible choice of $s_1, ..., s_m \in \{0,1\}^m$, the fraction of r's for which $R(x, r \oplus s_i) = 1$ holds for some i is at most $m \cdot \frac{1}{3m} = \frac{1}{3}$. Thus, the reduction always maps such an x to a NO-instance (i.e., an element of Π_{no}). On the other hand, we will show shortly that in case $\chi(x) = 1$, with probability at least $1/2$ the reduction maps x to a YES-instance. Thus, the above reduction has one-sided error and indeed reduces χ to Π (which as observed above is in co$\mathcal{RP}$). It is left to analyze the probability that the reduction fails in case $\chi(x) = 1$. That is,

$$\begin{aligned}
\Pr_{\bar{s}}[(x, \bar{s}) \notin \Pi_{\text{yes}}] &= \Pr_{s_1, ..., s_m}[\exists r \in \{0,1\}^m \text{ s.t. } (\forall i)\, R(x, r \oplus s_i) = 0] \\
&\leq \sum_{r \in \{0,1\}^m} \Pr_{s_1, ..., s_m}[(\forall i)\, R(x, r \oplus s_i) = 0] \\
&\leq 2^n \cdot \left(\frac{1}{3m}\right)^m \ll \frac{1}{2}
\end{aligned}$$

Comment: The traditional presentation uses the above reduction to show that $\mathcal{BPP}$ is in the Polynomial-Time Hierarchy. One defines the polynomial-time predicate $\varphi(x, \bar{s}, r) \overset{\text{def}}{=} \bigvee_{i=1}^{m}(R(x, s_i \oplus r) = 1)$, and observes that

$$\chi(x) = 1 \;\Rightarrow\; \exists \bar{s}\, \forall r\; \varphi(x, \bar{s}, r)$$
$$\chi(x) = 0 \;\Rightarrow\; \forall \bar{s}\, \exists r\; \neg\varphi(x, \bar{s}, r)$$

B.2.3 The Permanent: Worst-Case vs Average Case or, the Self-correction Paradigm

We consider the problem of computing the permanent of a matrix.[4] This problem is known to be $\#\mathcal{P}$-complete even in case the matrix has only 0-1 entries. Here we consider the problem of computing the permanent over sufficiently large finite fields (i.e., the field size is larger than the dimension). We show that the (worst-case) problem can be reduced to solving the problem on random (or typical) instances.

Reduction: On input an n-by-n matrix, M, over F (s.t., $|F| > n + 1$), the probabilistic polynomial-time oracle machine (i.e., the reduction) proceeds as follows.

1. Uniformly select an n-by-n matrix, R, over F.
2. For $i = 1, ..., n + 1$, obtain from the oracle the value, denoted v_i, of the permanent of the matrix $M + iR$.
3. Obtain by interpolation, the value of the degree n univariant polynomial, p, satisfying $p(i) = v_i$ (for $i = 1, ..., n + 1$).
4. Output $p(0)$.

The key observation, underlying the above reduction, is that, for fixed M and R, the permanent of $M + iR$ is a degree n polynomial in the variable i.

Analysis: We consider the performance of the above reduction assuming it is given access to an oracle which answers correctly on all but at most an $1/3(n + 1)$ fraction of the instances. We will show that in such a case, on any input, the reduction answers correctly with probability at least $2/3$. Observe that, for each fixed M and $i \neq 0$, the matrix $M + iR$ is uniformly distributed over the instance space. Thus, the probability that the oracle returns an incorrect answer on any of the $n + 1$ queries is at most $1/3$. But otherwise, having the permanent of $M + iR$ for every $i = 1, .., n + 1$, we obtain the permanent of the formal matrix $M + xR$ (which is a polynomial of degree n in $x \in$ F), and thus the permanent of M (when substituting $x = 0$).

---

[4] The permanent of an n-by-n matrix $A = (a_{i,j})$ is the sum, taken over all permutations π of $[n]$, of the products $\prod_{i=1}^{n} a_{i,\pi(i)}$.

Comments: As seen above, the reduction of a problem to random instances of itself allows to reduce its "worst" instances to its average (or typical) cases, and thus means that the problem does not really have "worst" (or "pathological") instances: The problem's complexity, in case the problem is hard, must stem from typical (or random) instances. Viewed from the other side (i.e., of feasibility), such a reduction allows to *self-correct* a procedure which is correct on a large majority of instances, and obtain a randomized procedure which is correct on every instance. Thus, as any reduction, a reduction to random instances is open to interpretation: For example, Ajtai's reduction of approximating shortest vectors in integer lattices to such random instances [3], is commonly viewed as a demonstration of average-case hardness based on worst-case hardness, but it may be also viewed as a self-corrector for programs which find short vectors in a certain class of integer lattices.

B.3 Randomness in Distributed Computing

As much as randomness is a powerful tool in the design of algorithms and reductions, its power in the distributed context is even more striking. In particular, randomized distributed protocols can beat impossibility results and lower bound which refer to deterministic protocols. Various examples are given in [100, 257, 22, 237, 269].

As a warm-up consider the problem of electing a leader among a set of n *identical* processes. Clearly, there is no deterministic procedure to elect such a leader (even when all processes are guaranteed to be non-faulty), as there is no way to "deterministically break the symmetry" among the processors. However, a simple randomized procedure will do the job: Let each processor toss, independently of all other processors, a coin with bias $1/n$ towards 1, and announce its coin-flip to all processors. If a single processor sends 1 then it is elected leader, otherwise the process is repeated. In general, randomness can be used to "break symmetry" in a variety of distributed settings. Other uses of randomness in such settings include avoiding "pathological" configurations (see Section B.3.2), and making the actions of non-faulty processors unpredictable to malicious ones (i.e., Byzantine faults; see Section B.3.3). We start with a much simpler problem.

B.3.1 Testing String Equality
or, Randomized Fingerprints

The problem considered here is to decide whether two strings, each held by a different party, are identical. The aim is to devise a protocol for this problem using low communication complexity. We present three such protocols.

Protocol 1: Party A holds $x \in \{0,1\}^n$, whereas party B holds $y \in \{0,1\}^n$. Here we view x, y as non-negative integers in $\{0, 1, ..., 2^n - 1\}$. In the protocol, party A uniformly selects $i \in \{1, ..., n\}$, finds the ith prime, denoted p_i, and sends the pair $(i, x \bmod p_i)$ to B. Party B recovers p_i and accepts if and only if $y \bmod p_i$ equals the value $x \bmod p_i$ (received from A).

Clearly, if $x = y$ then B always accepts. On the other hand, using the Chinese Reminder Theorem, we know that if $x \neq y$ then $x \neq y \pmod{p_i}$ for at least $n/2$ of the p_i's (or else $x \equiv y \pmod{\prod_{i \in I} p_i}$, for $|I| \geq n/2$, and $x = y$ follows as $x, y < 2^n < \prod_{i \in I} p_i$). The number of bits sent is $\log_2 n + \log_2(n \ln n)$.

Protocol 2: Again, party A holds $x \in \{0,1\}^n$, whereas party B holds $y \in \{0,1\}^n$. Here we use a small-bias probability space $S \subset \{0,1\}^n$, with bias $1/6$ and $|S| = \mathrm{poly}(n)$ (see Section 3.6.2). By definition, for every non-zero string $z \in \{0,1\}^n$, with probability at least $1/3$ a uniformly chosen $r \in S$ has inner product mod 2 with z equal to 1. In the protocol, party A uniformly selects $r \in S$, computes the inner product mod 2 of x and r, and sends the result along with the index of r (in S) to B. Party B retrieves r, computes the inner product mod 2 of y and r, and accepts if it matches the bit received.

Clearly, if $x = y$ then B always accepts. On the other hand, by the above, if $x \neq y$ then the inner products of x and y with a uniformly chosen $r \in S$ differ with probability at least $1/3$ (hint: consider $z = x \oplus y$). The number of bits sent is $1 + \log_2 |S| = O(\log n)$.

Protocol 3: The inputs are as above, but here we use a different tool: An error-correcting code, denoted $E : \{0,1\}^n \mapsto \{0,1\}^m$, with $m = O(n)$ and distance $\Omega(n)$ (cf., [225]). In the protocol, party A computes the codeword $E(x)$, uniformly selects $i \in \{1, ..., m\}$, and sends i along with the ith bit of $E(x)$ to Party B. The latter computes the codeword $E(y)$ and accepts if its ith bit matches the bit received.

Clearly, if $x = y$ then B always accepts. On the other hand, if $x \neq y$ then $E(x)$ and $E(y)$ differ on a constant fraction of the bit positions, and so B will reject with constant probability. The number of bits sent is $1 + \log_2 m = O(1) + \log_2 n$.

B.3.2 Routing in Networks
or, Avoiding Pathological Configurations

The problem considered here is to allow parallel routing of messages in a network in which processors have relatively few immediate neighbors (i.e., processors connected to them by a direct link). In many such networks, *routing to random destinations* can be done quite efficiently (i.e., fast even assuming that each processor can only deliver a single message at a time, and without coordination among the processors). Off course, we are interested in routing messages to "non-random" destinations; that is, to destinations which are imposed upon us by some high-level application. Still the above fact (regarding routing to random destinations) becomes relevant, via the following two phase *randomized routing* strategy: Suppose that processor i wishes to deliver a message to processor d_i, where the d_i's consist of an arbitrary a permutation of the processor names $[n] \stackrel{\text{def}}{=} \{1, ..., n\}$. Then, processor i selects a random intermediate processor, $r_i \in [n]$, and sends its message to processor r_i with a request to forward it to processor d_i. (The r_i's are not likely to be distinct!) Thus, the routing is in two phases:

1. The message of processor i, denoted m_i, is delivered to r_i.
2. Message m_i is delivered from r_i to d_i.

By our hypothesis, Phase 1 can be completed fast with high probability. It is appealing to say that, by symmetry, the same should hold also for Phase 2. This is not known to be generically true, but has been proved to be so for a wide class of networks (cf., [241, Sec. 3.4]). Specifically, if one changes the model a bit, allowing and measuring edge congestion, then bounds on congestion in Phase 1 apply also to Phase 2.

B.3.3 Byzantine Agreement
or, Take Actions the Adversary Cannot Predict

The problem considered here is to allow non-faulty processors to agree on a common value, in presence of Byzantine (malicious) faulty processors. Specifically, it is required that (1) the non-faulty processors must terminate with the same output value, and (2) in case their input values are the same this should also be their output value. We may consider, without loss of generality, the problem of agreeing on a Boolean value. The primary parameters are the total number of processors, denoted n, and a bound on the number of faulty processors, t. We assume a synchronous model of point-to-point communication.

Protocol: We use auxiliary (threshold) parameters L, H, D so that $L > \frac{n}{2} + t$, $H \geq L + t$ and $H + t \leq D \leq n - t$ (which is feasible for $t < n/8$). The protocol utilizes a *global coin* (which may be implemented in various ways). It is postulated that, for each flipping of this coin, each of the two possible outcomes occurs with probability at least $p > 0$ ($p = 0.1$ will do, whereas $p = 0.5$ corresponds to an unbiased coin).

Following is the program to processor $i \in [n] \stackrel{\text{def}}{=} \{1, ..., n\}$. On input $b_i \in \{0, 1\}$, the processor sets its (initial) vote, denoted vote_i, to b_i. The processor repeats the following steps $r + 1$ times, where r is the iteration in which it decides (see below):

1. Send vote_i to each processor.
2. Receive votes from all processors, including itself. (In case no message is received from processor j, use the value last received from it, and if no value was ever received use value 0.) Let cnt_i denote the number of votes

in favor of 1. If $\text{cnt}_i > n/2$ set $\text{maj}_i = 1$ and $\text{tally}_i = \text{cnt}_i$, otherwise set $\text{maj}_i = 0$ and $\text{tally}_i = n - \text{cnt}_i$.

3. Let $C \in \{L, H\}$ be the value of the global coin, for the current round (in each round the global coin is flipped anew).

4. If $\text{tally}_i \geq C$ then set $\text{vote}_i = \text{maj}_i$ else set $\text{vote}_i = 0$.

5. If $\text{tally}_i \geq D$ then *decide* vote_i, and proceed for a single additional iteration (skipping this step in the next iteration).

   (Actually, as shown below, if the processor were to decide again in the next iteration its decision would have been identical.)

Analysis: Let G denote the set of non-faulty (or good) processors. The following observation regarding members of G is extensively used: In each iteration, $|\text{cnt}_i - \text{cnt}_j| \leq t$, for every $i, j \in G$. Thus, if $\text{tally}_i \geq L > n/2 + t$ for some $i \in G$ then $\text{maj}_j = \text{maj}_i$ for all $j \in G$. Similarly, if $\text{tally}_i \geq D$ (resp., $\text{tally}_i \geq H$) for some $i \in G$ then $\text{tally}_j \geq H$ (resp., $\text{tally}_j \geq L$) for all $j \in G$. Using these facts it follows that

1. *If all good processors enter some round with identical votes then they all decide by the end of the current round, and their decision equals this vote.* This follows since (at this round) this identical vote would have support of at least $|G| \geq n - t \geq D$. (As a special case, we conclude that the second requirement of Byzantine Agreement holds.)

2. *If at some round a good processor decides v then by the end of the next round all good processors decide v.* Suppose that $i \in G$ decides v in the current round. Then, $\text{tally}_i \geq D$, and for each $j \in G$ it follows that $\text{tally}_j \geq H$ and so at Step 4 $\text{vote}_j = \text{maj}_j = v$. Using the previous fact, the current one follows. (As a special case, we conclude that the first requirement of Byzantine Agreement holds.)

3. *If at some round $\text{tally}_i \geq H$ holds for some $i \in G$ then with constant probability all good processors enter the next round with vote equal to maj_i.* This follows since with constant probability the outcome of the global coin is L, in which case for every $j \in G$, $\text{tally}_j \geq L = C$ and so at Step 4 $\text{vote}_j = \text{maj}_j = \text{maj}_i$.

4. *If at some round $\text{tally}_i < H$ holds for all $i \in G$ then with constant probability all good processors enter the next round with vote 0.* This follows since with constant probability the outcome of the global coin is H.

Thus, the above protocol terminates in *constant expected number of rounds*, and the output always satisfies the agreement requirements. This remain valid even if we use a global coin the outcome of which may be viewed differently by different processors, as long as for each of the two possible values, with probability at least $p > 0$, all non-faulty processors view the outcome as equal to that value. We comment that such a global coin can be easily implemented in case $t = O(\sqrt{n})$, by letting each processor toss a local coin, announce the outcome, and view the outcome of the global coin to be the majority vote it has received (which, with constant probability, will be identical at all good processors). We note that $t + 1$ is a lower bound on the number of rounds in any correct *deterministic* protocol. Furthermore, the above protocol can be adapted to the asynchronous model, whereas there exist no correct *deterministic* protocol for the latter model (even for $t = 1$).

B.4 Bibliographic Notes

Section B.1.1 (*approximating the number of DNF satisfying assignments*) is based on [229], Section B.1.2 (*finding perfect matching*) is based on [270], and Section B.1.3 (*testing polynomial identities*) is based on [321, 354]. The *Randomized Rounding* technique was introduced in [308], and the *MaxSAT application* described in Section B.1.4 is due to [163]. The *primality testing* algorithm described in Section B.1.5 is folklore attributed to several people; I heard it attributed to M. Blum. Section B.1.6 (*random walk algorithm for testing connectivity*) is based on [6], and Section B.1.7 (*the randomized min-cut algorithm*) is based on [227].

Section B.2.1 (*reduction of approximate counting to deciding* and of *SAT to uniqueSAT*) is based on [329, 334] and [344], but the presentation in these sources is quite different. The reduction of Section B.2.2 is based on [240], where it was used to show (independently of [329]) that $\mathcal{BPP} \in \mathcal{PH}$; the current presentation is due to Fortnow (priv. comm. 1997, see [14]). Section B.2.3 (*self-corrector for the permanent*) is based on [69, 248].

Protocol 1 for *string equality* (in Section B.3.1) is commonly attributed to M. Rabin and A. Yao, Protocol 2 is due to [274, Sec. 9], and Protocol 3 is due to E. Kushilevitz (priv. comm. 1998). Section B.3.2 (*randomized routing*) is based on [341, 343], and Section B.3.3 (*randomized Byzantine Agreement*) is based on [56, 305].

C. Two Proofs

In this appendix we provide proofs of two basic results. The first proof is for a folklore theorem which asserts that the soundness error in parallel repetition of interactive proofs deceases exponentially with the number of repetitions. To the best of our knowledge, a proof of this commonly utilized theorem has never appeared before. The proof itself is quite easy, but in light of the above we present it in full detail. The second proof provided in this appendix is for Theorem 3.11 asserting the existence of a generic hard-core predicate. This proof is different from the one which has appeared in the original text [182], and is provided in full detail (rather than in a terse form as in [182]).

C.1 Parallel Repetition of Interactive Proofs

By k parallel repetitions of an interactive proof system, (P, V), we mean a proof system (P_k, V_k) in which the parties play in parallel k copies of (P, V). That is, V_k (resp., P_k) generates k independently distributed random-pads, $r_1, ..., r_k$, for V (resp., $\omega_1, ..., \omega_k$ for P), and sets its ith message to $\beta_{1,i}, ..., \beta_{k,i}$, where $\beta_{j,i} = V(r_j, \alpha_{1,j}, ..., \alpha_{i-1,j})$ (resp., to $\alpha_{1,i}, ..., \alpha_{k,i}$, where $\alpha_{j,i} = P(\omega_j, \beta_{1,j}, ..., \beta_{i,j})$). We stress that V_k accepts if and only if V would have accepted in all k copies.[1] We are interested in the soundness error of V_k, which only depends on V and k (and so P_k and P are omitted from the rest of the discussion). For any pair of interactive machines, A and B, let use denote by (A, B) the output of A after interacting with B, on common input x. The Parallel Repetition Theorem for interactive proofs is captured by the following lemma.

Lemma C.1 (folklore): *Let V_1 be an interactive machine, and V_k be an interactive machine obtained from V_1 by playing k versions of V_1 in parallel. Let*

$$p_1(x) \overset{\text{def}}{=} \max_{P^*}\{\Pr[(P^*, V_1)(x) = 1]\}, \text{ and}$$

---

[1] The analysis of the case where V_k accepts iff a threshold number of copies accept is more complex; see [37]. The simple case treated here suffices for "error reduction" in interactive proofs with one-sided error. A threshold rule is typically employed when "reducing error" in two-sided error proof system.

$$p_k(x) \overset{\text{def}}{=} \max_{P^*}\{\Pr[(P^*, V_k)(x) = 1]\}.$$

Then

$$p_k(x) = p_1(x)^k$$

Proof: Clearly, $p_k(x) \geq p_1(x)^k$. The point is to prove $p_k(x) \leq p_1(x)^k$. We stress that one may not just assume that the optimal prover strategy against V_k consists of playing optimally but independently in each of the k parallel copies. As we shall see below, this conjecture turns out to be correct in the current setting (but is wrong in related settings such as multi-party interactive proofs and computationally-sound proofs; see [155, 135, 309, 136] and [46], respectively). Thus, a proof is due.

The proof uses the notion of the *game tree* of a proof system. Fixing a verifier V we consider its interaction with a generic prover on any fixed common input, denoted x. The verifier's random choices can be thought of as corresponding to the contents of its random-tape, called the random-pad. We assume without loss of generality that V sends the first message and that the prover sends the last one. In each round, V's message is chosen depending on the history of the interaction so far and according to some probability distribution induced by V's local random-tape. The history so far corresponds to a fixed subset of possible random-pads, and the possible messages to be sent correspond to a partition of this subset. Thus, each possible message is sent with probability proportional to its part in this subset. The above description corresponds to general interactive proofs. (In case of Arthur-Merlin games the situation is simpler: V merely tosses a predetermined number of coins and sends the outcome to the prover.) As to the prover's messages, they are chosen arbitrarily (but are of length at most poly($|x|$)). The interaction goes on, for at most poly($|x|$) rounds at which point the verifier stops outputting either *accept* or *reject*. The messages exchanged till that point are called a *transcript* of the interaction between the prover and V. To simplify the exposition, we augment the transcript of the interaction by V's random-pad. This way, V's accept/reject decision is determined by the *augmented transcript* (and the input x). The interaction between the prover and V on common input x may be viewed as a game in which the prover's objective is to maximize the probability that V accepts, and V's strategy is fixed but mixed (i.e., probabilistic).

Definition C.2 (the game tree and its value): *Let V and x be fixed.*

– *The tree T_x: The nodes in T_x correspond to prefixes of transcripts of possible interactions of V with an arbitrary prover. The root represents the empty interaction and is defined to be at level 0. For every $i \geq 0$, the edges going out from each $2i^{\text{th}}$ level node correspond to the messages V may send given the history so far. The edges going out from each $(2i + 1)^{\text{st}}$ level node correspond to the messages a prover may send given the history so far.*

Leaves correspond to augmented transcripts as defined above, and so their direct ancestors correspond to full transcripts.

- *The value of T_x:* *The value of the tree is defined bottom-up as follows. The* value of a leaf *is either 0 or 1 depending on whether V accepts in the augmented transcript represented by it or not. The* value of an internal node at level $2i$ *is defined as the weighted average of the values of its children, where the weights correspond to the probabilities of the various verifier messages.* (This definition holds also for the fathers of leaves, when viewing V's random-pad as an auxiliary, fictitious message sent by V.) *The* value of an internal node at level $2i-1$ *is defined as the maximum of the values of its children. This corresponds to the prover's strategy of trying to maximize V's accepting probability. The* value of the tree *is defined as the value of its root.*

We may assume, without loss of generality, that the averages taken in even-leveled nodes are plain averages (rather than weighted ones). This is justified by duplicating odd-level nodes. We stress that this modification is applied to the game-tree (not to the verifier), and results in a tree the correspondence of which to the proof system is less obvious. Notice that we are dealing with a general interactive proof, yet our analysis of the game-tree is a mental experiment (which need not be efficiently implementable).

We consider the game-trees of both the basic proof system and the k-repeated proof system. Fixing an input, we denote the first tree by T_1 and the second by T_k. There is a natural 1-1 mapping of nodes in T_k to sequences of k nodes in T_1. Going from the leaves of T_k to its root, we prove by induction that the value of each node is T_k equals the product of the values of the k nodes to which it is mapped (by the above mapping). The base case (i.e., leaves) is obvious, and there are two cases to consider in the induction step.

1. For a prover-node, $\mathbf{v} = (v_1, ..., v_k)$, denote its children in T_k by $\mathbf{w^i} = (w_1^{i_1}, ..., w_k^{i_k})$, where $\mathbf{i} = (i_1, ..., i_k)$ and w_j^i is the i-th child in T_1 of v_j. Then, by definition of the game trees

$$\text{val}(\mathbf{v}) = \max_{\mathbf{i}}(\text{val}(\mathbf{w^i})), \text{ and} \tag{C.1}$$

$$\text{val}(v_j) = \max_{i}(\text{val}(w_j^i)), \text{ for } j = 1, ..., k. \tag{C.2}$$

By induction, for every $\mathbf{i} = (i_1, ..., i_k)$,

$$\text{val}(\mathbf{w^i}) = \prod_{j=1}^{k} \text{val}(w_j^{i_j})) \tag{C.3}$$

Combining Equations (C.1)–(C.3), and using the "distributive feature" of maximization, we get

$$\mathrm{val}(\mathbf{v}) \;=\; \max_{\mathbf{i}}(\mathrm{val}(\mathbf{w}^i))$$

$$=\; \max_{\mathbf{i}}\left(\prod_{j=1}^{k}\mathrm{val}(w_j^{i_j})\right)$$

$$=\; \prod_{j=1}^{k}\max_{i_j}(\mathrm{val}(w_j^{i_j}))$$

$$=\; \prod_{j=1}^{k}\mathrm{val}(v_j)$$

as required.

2. For a verifier-node, $\mathbf{v} = (v_1, ..., v_k)$, denote its children in T_k by $\mathbf{w^i} = (w_1^{i_1}, ..., w_k^{i_k})$, where $\mathbf{i}$ and the w_j^i's are as above. Then, by definition of the game trees

$$\mathrm{val}(\mathbf{v}) \;=\; \mathrm{aver}_{\mathbf{i}}(\mathrm{val}(\mathbf{w}^i)), \quad \text{and} \tag{C.4}$$

$$\mathrm{val}(v_j) \;=\; \mathrm{aver}_i(\mathrm{val}(w_j^i)), \quad \text{for } j = 1, ..., k. \tag{C.5}$$

where $\mathrm{aver}_i(x_i)$ denotes the average value of the x_i's which are to be understood from the context. Again, Eq. (C.3) holds by induction, and so we get

$$\mathrm{val}(\mathbf{v}) \;=\; \mathrm{aver}_{\mathbf{i}}(\mathrm{val}(\mathbf{w}^i))$$

$$=\; \mathrm{aver}_{\mathbf{i}}\left(\prod_{j=1}^{k}\mathrm{val}(w_j^{i_j})\right)$$

$$=\; \prod_{j=1}^{k}\mathrm{aver}_{i_j}(\mathrm{val}(w_j^{i_j}))$$

$$=\; \prod_{j=1}^{k}\mathrm{val}(v_j)$$

The lemma follows.   □

We comment that the above argument generalizes to the case in which the k copies of V_1 are invoked on possibly different inputs. That is,

Lemma C.3 *Let V_1 be an interactive machine, and V_k be an interactive machine obtained from V_1 by playing k versions of V_1 in parallel so that on input $\overline{x} = (x_1, ..., x_k)$ to V_k the ith version of V_1 is invoked on x_i. Let $p_1(x) \stackrel{\mathrm{def}}{=} \max_{P^*}\{\Pr[(P^*, V_1)(x) = 1]\}$, and $p_k(\overline{x}) \stackrel{\mathrm{def}}{=} \max_{P^*}\{\Pr[(P^*, V_k)(\overline{x}) = 1]\}$. Then*

$$p_k(x_1, ..., x_k) \;=\; \prod_{i=1}^{k}p_1(x_i)$$

Perspective – parallel repetition in multi-prover interactive proofs.
To demonstrate the dependency of the above lemma on the full-information
setting of interactive proof systems, we reproduce a counter-example to the
analogous claim for two-prover proof systems. (The counter-example is due
to Feige [135], improving over [155].) The basic one-round two-prover system
is as follows.

1. The verifier uniformly selects two bits $b_1, b_2 \in \{0, 1\}$, and sends b_1 (resp.,
   b_2) to the first (resp., second) prover.
2. Each prover is supposed to reply with a pair $(i, r) \in \{1, 2\} \times \{0, 1\}$.
3. Upon receiving (i_1, r_1) and (i_2, r_2), from the first and second prover respectively, the verifier accepts if and only if $(i_1, r_1) = (i_2, r_2)$ and $r_1 = b_{i_1}$.
   (That is, both provers should reply with the identity of one of the provers
   and the bit sent to it.)

It can be easily shown that the value of this basic game is $1/2$. We care
only about the upper bound which is established by noting that in order to
win both provers must send the same message but only one of them knows
the relevant bit of the verifier.[2] We now consider the parallel execution of
two copies of the basic game. For clarity, we explicitly present the resulting
parallel game.

1. The verifier uniformly selects four bits $b_1^1, b_2^1, b_1^2, b_2^2 \in \{0, 1\}$, and sends
   (b_1^1, b_1^2) (resp., (b_2^1, b_2^2)) to the first (resp., second) prover.
2. Each prover is supposed to reply with two pair $(i^1, r^1), (i^2, r^2) \in \{1, 2\} \times \{0, 1\}$.
3. Upon receiving $((i_1^1, r_1^1), (i_1^2, r_1^2))$ and $((i_2^1, r_2^1), (i_2^2, r_2^2))$, from the first and
   second prover respectively, the verifier accepts if and only if $(i_1^j, r_1^j) = (i_2^j, r_2^j)$ and $r_1^j = b_{i_1^j}^j$, for both $j = 1, 2$.

It can be shown that the value of this parallel game remains $1/2$, thus providing a dramatic refutation to the naive parallel repetition conjecture for
multi-prover proof systems. In particular, we note that the provers can win
with probability $1/2$ if the first (resp., second) prover, upon receiving (b_1^1, b_1^2)
(resp., (b_2^1, b_2^2)), respond with $((1, b_1^1), (2, b_1^1))$ (resp., $((1, b_2^2), (2, b_2^2))$). The reason being that these strategies win if and only if $b_1^1 = b_2^2$ (which happens
with probability $1/2$).

C.2 A Generic Hard-Core Predicate

Theorem 3.11, due to Goldreich and Levin [182], relates two computational
tasks: The first task is inverting a function f; namely given y find an x so that
$f(x) = y$. The second task is predicting, with non-negligible advantage, the
exclusive-or of a subset of the bits of x when only given $f(x)$. More precisely,

---

[2] The lower bound may be established by both provers always replying $(1, 0)$.

it has been proved that if f cannot be efficiently inverted then given $f(x)$ and r it is infeasible to predict the inner-product mod 2 of x and r better than obvious.

The proof presented here is not the original one presented in [182] (see generalization in [190]), but rather an alternative suggested by Charlie Rackoff. The alternative proof, inspired by [7], has two main advantages over the original one: It is simpler to explain, and it provides better security (i.e., a more efficient reduction of inverting f to predicting the inner-product).

Theorem C.4 (Theorem 3.11 – restated): *Let $b(x, r)$ denote the inner-product mod 2 of the binary vectors x and r. Suppose we have oracle access to a random process $b_x : \{0, 1\}^n \mapsto \{0, 1\}$, so that*

$$\Pr_{r \in \{0,1\}^n}[b_x(r) = b(x, r)] \geq \frac{1}{2} + \epsilon$$

where the probability is taken uniformly over the internal coin tosses of b_x and all possible choices of $r \in \{0, 1\}^n$. Then we can output, in time polynomial in n/ϵ, a list of strings which with probability at least $\frac{1}{2}$ contains x.

Theorem 3.11 is derived from the above by using standard arguments. We prove this fact first.

Proposition C.5 *Theorem C.4 implies Theorem 3.11.*

Proof: We assume for contradiction the existence of an efficient algorithm predicting the inner-product with advantage which is not negligible, and derive an algorithm that inverts f with related (i.e., not negligible) success probability. This contradicts the hypothesis that f is a one-way function. Thus, the proof uses a "reducibility argument" – that is, we reduce the task of inverting f to the task of predicting $b(x, r)$ from $(f(x), r)$.

Let G be a (probabilistic polynomial-time) algorithm that on input $f(x)$ and r tries to predict the inner-product (mod 2) of x and r. Denote by $\epsilon_G(n)$ the (overall) advantage of algorithm G in predicting $b(x, r)$ from $f(x)$ and r, where x and r are uniformly chosen in $\{0, 1\}^n$. Namely,

$$\epsilon_G(n) \overset{\text{def}}{=} \Pr[G(f(X_n), R_n) = b(X_n, R_n)] - \frac{1}{2}$$

where here and in the sequel X_n and R_n denote two independent random variables, each uniformly distributed over $\{0, 1\}^n$. In the sequel we shorthand ϵ_G by ϵ.

Our first observation is that, on at least an $\frac{\epsilon(n)}{2}$ fraction of the x's of length n, algorithm G has an $\frac{\epsilon(n)}{2}$ advantage in predicting $b(x, R_n)$ from $f(x)$ and R_n. Namely,

Claim: There exists a set $S_n \subseteq \{0, 1\}^n$ of cardinality at least $\frac{\epsilon(n)}{2} \cdot 2^n$ such that for every $x \in S_n$, it holds that

$$s(x) \stackrel{\text{def}}{=} \Pr[G(f(x), R_n) = b(x, R_n)] \geq \frac{1}{2} + \frac{\epsilon(n)}{2}$$

This time the probability is taken over all possible values of R_n and all internal coin tosses of algorithm G, whereas x is fixed.

Proof: The observation follows by an averaging argument. Namely, write $\text{Exp}(s(X_n)) = \frac{1}{2} + \epsilon(n)$, and apply Markov Inequality.$\square$

Thus, we restrict our attention to x's in S_n. For each such x, the conditions of Theorem C.4 hold, and so within time $\text{poly}(n/\epsilon(n))$ and with probability at least $1/2$ we retrieve a list of strings containing x. Contradiction to the one-wayness of f follows, since the probability we invert f on uniformly selected x is at least $\frac{1}{2} \cdot \Pr[U_n \in S_n] \geq \frac{\epsilon(n)}{4}$. $\square$

C.2.1 A Motivating Discussion

Let $s(x) \stackrel{\text{def}}{=} \Pr[b_x(r) = b(x, r)]$, where r is uniformly distributed in $\{0, 1\}^{|x|}$. Then, by the hypothesis of Theorem C.4, $s(x) \geq \frac{1}{2} + \epsilon$. Suppose, for a moment, that $s(x) > \frac{3}{4} + \epsilon$. In this case, retrieving x by querying the oracle b_x is quite easy. To retrieve the ith bit of x, denoted x_i, we uniformly select $r \in \{0, 1\}^n$, and obtain $b_x(r)$ and $b_x(r \oplus e^i)$, where e^i is an n-dimensional binary vector with 1 in the ith component and 0 in all the others, and $v \oplus u$ denotes the addition mod 2 of the binary vectors v and u. Clearly, if both $b_x(r) = b(x, r)$ and $b_x(r \oplus e^i) = b(x, r \oplus e^i)$ then

$$\begin{aligned} b_x(r) \oplus b_x(r \oplus e^i) &= b(x, r) \oplus b(x, r \oplus e^i) \\ &= b(x, e^i) \\ &= x_i \end{aligned}$$

The probability that both equalities hold (i.e., both $b_x(r) = b(x, r)$ and $b_x(r \oplus e^i) = b(x, r \oplus e^i)$) is at least $1 - 2 \cdot (\frac{1}{4} - \epsilon) = \frac{1}{2} + 2\epsilon$. Hence, repeating the above procedure sufficiently many times and ruling by majority we retrieve x_i with very high probability. Similarly, we can retrieve all the bits of x, and hence obtain x itself. However, the entire analysis was conducted under (the unjustifiable) assumption that $s(x) > \frac{3}{4} + \epsilon$, whereas we only know that $s(x) > \frac{1}{2} + \epsilon$.

The problem with the above procedure is that it doubles the original error probability of the oracle b_x on random queries. Under the unrealistic assumption, that the b_x's error on such inputs is significantly smaller than $\frac{1}{4}$, the "error-doubling" phenomenon raises no problems. However, in general (and even in the special case where b_x's error is exactly $\frac{1}{4}$) the above procedure is unlikely to yield x. Note that the error probability of b_x can not be decreased by querying b_x several times on the same instance (e.g., b_x may always answer correctly on three quarters of the inputs, and always err on the remaining quarter). What is required is an *alternative way of using b_x* –

a way which does not double the original error probability of b_x. The key idea is to generate the r's in a way which requires querying b_x only once per each r (and x_i), instead of twice. The good news are that the error probability is no longer doubled, since we will only use b_x to get an "estimate" of $b(x, r \oplus e^i)$. The bad news are that we still need to know $b(x, r)$, and it is not clear how we can know $b(x, r)$ without querying b_x. The answer is that we can guess $b(x, r)$ by ourselves. This is fine if we only need to guess $b(x, r)$ for one r (or logarithmically in $|x|$ many r's), but the problem is that we need to know (and hence guess) $b(x, r)$ for polynomially many r's. An obvious way of guessing these $b(x, r)$'s yields an exponentially vanishing success probability. The solution is to generate these polynomially many r's so that, on one hand they are "sufficiently random" whereas on the other hand we can guess all the $b(x, r)$'s with non-negligible success probability. Specifically, generating the r's in a *particular* **pairwise independent** manner will satisfy both (seemingly contradictory) requirements. We stress that in case we are successful (in our guesses for the $b(x, r)$'s), we can retrieve x with high probability. Hence, we retrieve x with non-negligible probability.

A word about the way in which the pairwise independent r's are generated (and the corresponding $b(x, r)$'s are guessed) is indeed in place. To generate $m = \text{poly}(n/\epsilon)$ many r's, we uniformly (and independently) select $l \stackrel{\text{def}}{=} \log_2(m+1)$ strings in $\{0, 1\}^n$. Let us denote these strings by $s^1, ..., s^l$. We then guess $b(x, s^1)$ through $b(x, s^l)$. Let us denote these guesses, which are uniformly (and independently) chosen in $\{0, 1\}$, by σ^1 through σ^l. Hence, the probability that all our guesses for the $b(x, s^i)$'s are correct is $2^{-l} = \frac{1}{\text{poly}(n/\epsilon)}$. The different r's correspond to the different non-empty subsets of $\{1, 2, ..., l\}$. We compute $r^J \stackrel{\text{def}}{=} \bigoplus_{j \in J} s^j$. The reader can easily verify that the r^J's are pairwise independent and each is uniformly distributed in $\{0, 1\}^n$. The key observation is that

$$b(x, r^J) = b\left(x, \bigoplus_{j \in J} s^j\right) = \bigoplus_{j \in J} b(x, s^j)$$

Hence, our guess for the $b(x, r^J)$'s is $\bigoplus_{j \in J} \sigma^j$, and with non-negligible probability all our guesses are correct.

C.2.2 Back to the Formal Argument

Following is a formal description of the recovering algorithm, denoted A. On input n and ϵ (and oracle access to b_x), algorithm A sets $l \stackrel{\text{def}}{=} \lceil \log_2(n \cdot \epsilon^{-2} + 1) \rceil$. Algorithm A uniformly and independently select $s^1, ..., s^l \in \{0, 1\}^n$, and $\sigma^1, ..., \sigma^l \in \{0, 1\}$. It then computes, for every non-empty set $J \subseteq \{1, 2, ..., l\}$, a string $r^J \leftarrow \bigoplus_{j \in J} s^j$ and a bit $\rho^J \leftarrow \bigoplus_{j \in J} \sigma^j$. For every $i \in \{1, ..., n\}$ and every *non-empty* $J \subseteq \{1, .., l\}$, algorithm A computes $z_i^J \leftarrow \rho^J \oplus b_x(r^J \oplus e^i)$. Finally, algorithm A sets z_i to be the majority of the z_i^J values, and outputs $z = z_1 \cdots z_n$.

Comment: An alternative implementation of the above ideas results in an algorithm, denoted A', which fits the conclusion of the theorem. Rather than selecting at random a setting of $\sigma^1, ..., \sigma^l \in \{0, 1\}$, algorithm A' tries all possible values for $\sigma^1, ..., \sigma^l$. It outputs a list of 2^l candidates z's, one per each of the possible settings of $\sigma^1, ..., \sigma^l \in \{0, 1\}$.

Clearly, A makes $n \cdot 2^l = n^2/\epsilon^2$ oracle calls to b_x, and the same amount of other elementary computations. Algorithm A' makes the same queries, but conducts a total of $(n/\epsilon^2) \cdot (n^2/\epsilon^2)$ elementary computations.

Following is a detailed analysis of the success probability of algorithm A. We start by showing that, in case the σ^j's are correct, then with constant probability, $z_i = x_i$ for all $i \in \{1, ..., n\}$. This is proven by bounding from below the probability that the majority of the z_i^J's equals x_i.

Claim: For every $1 \leq i \leq n$,

$$\Pr\left[|\{J : b(x, r^J) \oplus b_x(r^J \oplus e^i) = x_i\}| > \frac{1}{2} \cdot (2^l - 1)\right] > 1 - \frac{1}{4n}$$

where $r^J \stackrel{\text{def}}{=} \bigoplus_{j \in J} s^j$ and the s^j's are independently and uniformly chosen in $\{0, 1\}^n$.

Proof: For every J, define a 0-1 random variable ζ^J, so that ζ^J equals 1 if and only if $b(x, r^J) \oplus b_x(r^J \oplus e^i) = x_i$. The reader can easily verify that each r^J is uniformly distributed in $\{0, 1\}^n$. It follows that each ζ^J equals 1 with probability $\frac{1}{2} + \epsilon$. We show that the ζ^J's are pairwise independent by showing that the r^J's are pairwise independent. For every $J \neq K$ we have, without loss of generality, $j \in J$ and $k \in K \setminus J$. Hence, for every $\alpha, \beta \in \{0, 1\}^n$, we have

$$\begin{aligned} \Pr\left[r^K = \beta \mid r^J = \alpha\right] &= \Pr\left[s^k = \beta \mid s^j = \alpha\right] \\ &= \Pr\left[s^k = \beta\right] \\ &= \Pr\left[r^K = \beta\right] \end{aligned}$$

and pairwise independence of the r^J's follows. Let $m \stackrel{\text{def}}{=} 2^l - 1$. Using Chebyshev's Inequality, we get

$$\begin{aligned} \Pr\left[\sum_J \zeta^J \leq \frac{1}{2} \cdot m\right] &\leq \Pr\left[\left|\sum_J \zeta^J - (0.5 + \epsilon) \cdot m\right| \geq \epsilon \cdot m\right] \\ &< \frac{\text{Var}(\zeta^{\{1\}})}{\epsilon^{-2} \cdot (n/\epsilon^2)} \\ &< \frac{1}{4n} \end{aligned}$$

The claim now follows. $\square$

Recall that if $\sigma^j = b(x, s^j)$, for all j's, then $\rho^J = b(x, r^J)$ for all non-empty J's. In this case z output by algorithm A equals x, with probability at least $3/4$. However, the first event happens with probability $2^{-l} = \frac{1}{n/\epsilon^2}$ independently of the events analyzed in the Claim. Hence, algorithm A recovers x with probability at least $\frac{3}{4} \cdot \frac{\epsilon^2}{n}$ (whereas, the modified algorithm, A', succeeds with probability at least $\frac{3}{4}$). Theorem C.4 follows.   □

C.2.3 Improved Implementation of Algorithm A'

In continuation to the proof of Theorem C.4, we present guidelines for a more efficient implementation of Algorithm A'. In the sequel it will be more convenient to use arithmetic of reals instead of that of Boolean values. Hence, we denote $b'(x, r) = (-1)^{b(r,x)}$ and $b'_x(r) = (-1)^{b_x(r)}$.

1. Prove that $\mathrm{Exp}_r(b'(x,r) \cdot b'_x(r + e^i)) = 2\epsilon \cdot (-1)^{x_i}$, where $\epsilon = \mathrm{Pr}_r(b_x(r) = b(x,r)) - 0.5$.

2. Let v be an l-dimensional Boolean vector, and let R be a uniformly chosen l-by-n Boolean matrix. Prove that for every $v \neq u \in \{0,1\}^l \setminus \{0\}^l$ it holds that vR and uR are pairwise independent and uniformly distributed in $\{0,1\}^n$.

   (Each such vR corresponds to a r^J above, with $J = \{j : v_j = 1\}$.)

3. Prove that, with probability at least $\frac{1}{2}$ over the choices of R, there exists $u \in \{0,1\}^l$ so that, for every $1 \leq i \leq n$, the sign of $\sum_{v \in \{0,1\}^l} b'(u,v) b'_x(vR + e^i)$ equals the sign of $(-1)^{x_i}$.

   (Hint: Re-do the proof of the Claim of subsection C.2.2, using $b'(x, vR) = b'(xR^T, v)$ and $u \stackrel{\mathrm{def}}{=} xR^T$.)

4. Let B be a fixed 2^l-by-2^l matrix with the (u, v)-entry being $b'(u,v)$, and denote by $\bar{o}^i$ an 2^l-dimensional vector with the vth entry equal $b'_x(vR + e^i)$. Then, $B\bar{o}^i$ is an 2^l-dimensional vector with the uth entry equal to $\sum_{v \in \{0,1\}^l} b'(u,v) b'_x(vR + e^i)$.

   Consider an algorithm that uniformly selects an l-by-n matrix R, computes $\bar{y}_i \leftarrow B\bar{o}^i$, for all i's, and forms a 2^l-by-n matrix Y in which the columns are the $\bar{y}_i$'s. Let Z be a corresponding matrix in which the (u, i)-entry is 0 if the (u, i)-entry of Y is positive, and is 1 otherwise. The output is the list of rows in Z.

   (Notice that the algorithm makes $2^l \cdot n$ queries to obtain all entries in the $\bar{o}^i$'s, that all these queries can be computed within $2^l n$ time, and so all that remains is to multiply the fixed matrix B by the n vectors, $\bar{o}^i$'s.)

   a) Using Item 3, evaluate the success probability of the algorithm (i.e., the probability that x is in the output list).

   b) Using the special structure of the fixed matrix B, show that the product $B\bar{o}^i$ can be computed in time $l \cdot 2^l$.

      Hint: B is the Sylvester matrix, which can be written recursively as

$$S_k = \left(\begin{array}{cc} S_{k-1} & S_{k-1} \\ S_{k-1} & \overline{S_{k-1}} \end{array} \right)$$

where $S_0 = +1$ and $\overline{M}$ means flipping the $+1$ entries of M to -1 and vice versa. (Alternatively, note that $B\overline{o}$ is the Discrete Fourier Transform of $\overline{o}$.)

It follows that algorithm A' can be implemented in time $n \cdot l2^l$, which is $\widetilde{O}(n^2/\epsilon^2)$.

Further Improvement. We may further improve algorithm A' by observing that it suffices to let $2^l = O(1/\epsilon^2)$ rather than $2^l = O(n/\epsilon^2)$. Under the new setting, with constant probability, we recover correctly a constant fraction of the bits of x rather than all of them. If x were an codeword under an asymptotically good error-correcting code (cf., [225]), this would suffice. To avoid this assumption, we modify algorithm A' so that it tries to recover certain XORs of bits of x (rather than individual bits of x). Specifically, we use an asymptotically good linear code (i.e., having constant rate, correcting a constant fraction of errors and having efficient decoding algorithm) [225]. Thus, the modified A' recovers correctly a constant fraction of the bits in the encoding of x, under such codes, and using the decoding algorithm – recovers x.

D. Related Surveys by the Author

Reproduced below are abstracts of other surveys of the author on topics related to Randomness and Computation. All these surveys are available from the ECCC, the *Electronic Colloquium on Computational Complexity*, accessible by URL http://www.eccc.uni-trier.de/eccc/.

On Yao's XOR-Lemma [with N. Nisan and A. Wigderson, *ECCC*, TR95-050, 1995]: A fundamental lemma of Yao states that computational weak-unpredictability of functions gets amplified if the results of several independent instances are XORed together. We survey two known proofs of Yao's Lemma, and present a third alternative proof. The third proof proceeds by first proving that a function constructed by concatenating the values of the function on several independent instances is much more unpredictable, with respect to specified complexity bounds, than the original function. This statement turns out to be easier to prove than Yao's XOR-Lemma. Using a result of Goldreich and Levin, and some elementary observation, we derive Yao's XOR-Lemma.

Three XOR-Lemmas – An Exposition [*ECCC*, TR95-056, 1995]: We provide an exposition of three lemmas which relate general properties of distributions with the exclusive-or of certain bit locations. The first XOR-Lemma, commonly attributed to U.V. Vazirani, relates the statistical distance of a distribution from uniform to the maximum bias of the xor of certain bit positions. The second XOR-Lemma, due to U.V. Vazirani and V.V. Vazirani, is a computational analogue of the first: It relates the pseudorandomness of a distribution to the difficulty of predicting the xor of bits in particular (or random) positions. The third Lemma, due to Goldreich and Levin, relates the difficulty of retrieving a string and the unpredictability of the xor of random bit positions. The most notable XOR-Lemma – that is the so-called Yao XOR-Lemma is not discussed here.

A Sample of Samplers – A Computational Perspective on Sampling [*ECCC*, TR97-020, 1997]: We consider the problem of estimating the average of a huge set of values. That is, given oracle access to an arbitrary function $f : \{0,1\}^n \mapsto [0,1]$, we need to estimate $2^{-n} \sum_{x \in \{0,1\}^n} f(x)$ upto an additive error of ϵ. We are allowed to employ a randomized algorithm which may err with probability at most δ. We discuss lower and upper bounds, algorithms and the ideas underlying their construction, culminating in the best algorithm known. This algorithm makes $O(\epsilon^{-2} \cdot \log(1/\delta))$ queries and uses $n + O(\log(1/\epsilon)) + O(\log(1/\delta))$ coin tosses, both complexities being very close to the corresponding lower bounds.

Combinatorial Property Testing – A Survey [*ECCC*, TR97-056, 1997]: We consider the question of determining whether a given object has a pre-determined property or is "far" from any object having the property. Specifically, objects are modeled by functions, and distance between functions is measured as the fraction of the domain on which the functions differ. We consider (randomized) algorithms which may query the function at arguments of their choice, and seek algorithms which query the function at relatively few places. We focus on combinatorial properties, and specifically on graph properties. The two standard representations of graphs – by adjacency matrices and by incidence lists – yield two different models for testing graph properties. In the first model, most appropriate for dense graphs, distance between N-vertex graphs is measured as the fraction of edges on which the graphs disagree over N^2. In the second model, most appropriate for bounded-degree graphs, distance between N-vertex d-degree graphs is measured as the fraction of edges on which the graphs disagree over dN. To illustrate the two models, we survey results regarding the complexity of testing whether a graph is Bipartite. For a constant distance parameter, a constant number of queries suffice in the first model, whereas $\widetilde{\Theta}(\sqrt{N})$ queries are necessary and sufficient in the second model.

Notes on Levin's Theory of Average-Case Complexity [*ECCC*, TR97-058, 1997]: In 1984, Leonid Levin has initiated a theory of average-case complexity. We provide an exposition of the basic definitions suggested by Levin, and discuss some of the considerations underlying these definitions.

Bibliography

1. W. Aiello, M. Bellare and R. Venkatesan. Knowledge on the Average – Perfect, Statistical and Logarithmic. In *27th ACM Symposium on the Theory of Computing*, pages 469–478, 1995.
2. W. Aiello and J. Håstad. Perfect Zero-Knowledge Languages can be Recognized in Two Rounds. In *28th IEEE Symposium on Foundations of Computer Science*, pages 439–448, 1987.
3. M. Ajtai. Generating Hard Instances of Lattice Problems. In *28th ACM Symposium on the Theory of Computing*, pages 99–108, 1996.
4. M. Ajtai, J. Komlos, E. Szemerédi. Deterministic Simulation in LogSpace. In *19th ACM Symposium on the Theory of Computing*, pages 132–140, 1987.
5. M. Ajtai and A. Wigderson. Deterministic simulation of probabilistic constant depth circuits. In *26th IEEE Symposium on Foundations of Computer Science*, pages 11–19, 1985.
6. R. Aleliunas, R.M. Karp, R.J. Lipton, L. Lovász and C. Rackoff. Random walks, universal traversal sequences, and the complexity of maze problems. In *20th IEEE Symposium on Foundations of Computer Science*, pages 218–223, 1979.
7. W. Alexi, B. Chor, O. Goldreich and C.P. Schnorr. RSA/Rabin Functions: Certain Parts are As Hard As the Whole. *SIAM Journal on Computing*, Vol. 17, April 1988, pages 194–209.
8. N. Alon. Eigenvalues and expanders. *Combinatorica*, Vol. 6, pages 83–96, 1986.
9. N. Alon, L. Babai and A. Itai. A fast and Simple Randomized Algorithm for the Maximal Independent Set Problem. *J. of Algorithms*, Vol. 7, pages 567–583, 1986.
10. N. Alon, J. Bruck, J. Naor, M. Naor and R. Roth. Construction of Asymptotically Good, Low-Rate Error-Correcting Codes through Pseudo-Random Graphs. *IEEE Transactions on Information Theory*, Vol. 38, pages 509–516, 1992.
11. N. Alon, O. Goldreich, J. Håstad, R. Peralta. Simple Constructions of Almost k-wise Independent Random Variables. *Journal of Random structures and Algorithms*, Vol. 3, No. 3, (1992), pages 289–304.
12. N. Alon and V.D. Milman. λ_1, Isoperimetric Inequalities for Graphs and Superconcentrators, *J. Combinatorial Theory, Ser. B*, Vol. 38, pages 73–88, 1985.
13. N. Alon and J.H. Spencer. *The Probabilistic Method*, John Wiley & Sons, Inc., 1992.
14. A.E. Andreev, A.E.F. Clementi, J.D.P. Rolin and L. Trevisan, Weak Random Sources, Hitting Sets, and BPP Simulations. To appear in *SIAM Journal on Computing*. Preliminary version in *38th IEEE Symposium on Foundations of Computer Science*, pages 264–272, 1997.

15. R. Armoni, M. Saks, A. Wigderson and S. Zhou. Discrepancy sets and pseudorandom generators for combinatorial rectangles. In *37th IEEE Symposium on Foundations of Computer Science*, pages 412-421, 1996.

16. R. Armoni, A. Ta-Shma, A. Wigderson and S. Zhou. $SL \subseteq L^{4/3}$. In *29th ACM Symposium on the Theory of Computing*, pages 230–239, 1997.

17. R. Armoni and A. Wigderson. Pseudorandomness for space-bounbded computation. Unpublished manuscript, 1995.

18. S. Arora and C. Lund. Hardness of Approximations. In *Approximation Algorithms for NP-hard Problems*, D. Hochbaum ed., PWS, 1996.

19. S. Arora, C. Lund, R. Motwani, M. Sudan and M. Szegedy. Proof Verification and Intractability of Approximation Problems. In *33rd IEEE Symposium on Foundations of Computer Science*, pages 14–23, 1992.

20. S. Arora and S. Safra. Probabilistic Checkable Proofs: A New Characterization of NP. In *33rd IEEE Symposium on Foundations of Computer Science*, pages 1–13, 1992.

21. S. Arora and S. Sudan. Improved low degree testing and its applications. In *29th ACM Symposium on the Theory of Computing*, pages 485–495, 1997.

22. H. Attiya and J. Welch. *Distributed Computing: Fundamentals, Simulations and Advanced Topics*. McGraw-Hill Publishing Company, London, 1998.

23. L. Babai. Trading Group Theory for Randomness. In *17th ACM Symposium on the Theory of Computing*, pages 421–429, 1985.

24. L. Babai, L. Fortnow, and C. Lund. Non-Deterministic Exponential Time has Two-Prover Interactive Protocols. *Computational Complexity*, Vol. 1, No. 1, pages 3–40, 1991. Preliminary version in *31st IEEE Symposium on Foundations of Computer Science*, 1990.

25. L. Babai, L. Fortnow, L. Levin, and M. Szegedy. Checking Computations in Polylogarithmic Time. In *23rd ACM Symposium on the Theory of Computing*, pages 21–31, 1991.

26. L. Babai, L. Fortnow, N. Nisan and A. Wigderson. BPP has Subexponential Time Simulations unless EXPTIME has Publishable Proofs. *Complexity Theory*, Vol. 3, pages 307–318, 1993.

27. L. Babai and S. Moran. Arthur-Merlin Games: A Randomized Proof System and a Hierarchy of Complexity Classes. *Journal of Computer and System Science*, Vol. 36, pp. 254–276, 1988.

28. L. Babai, N. Nisan and M. Szegedy. Multiparty protocols, pseudorandom generators for logspace, and time-space trade-offs. *Journal of Computer and System Science*, Vol. 45(2), pgaes 204–232, 1992.

29. E. Bach and J. Shallit. *Algorithmic Number Theory* (Volume I: Efficient Algorithms). MIT Press, 1996.

30. D. Beaver. Foundations of Secure Interactive Computing. In *Crypto91*, Springer-Verlag Lecture Notes in Computer Science (Vol. 576), pages 377–391.

31. D. Beaver and J. Feigenbaum. Hiding Instances in Multioracle Queries. In *7th STACS*, Springer Verlag, Lecture Notes in Computer Science (Vol. 415), pages 37–48, 1990.

32. M. Bellare, R. Canetti and H. Krawczyk. Pseudorandom functions Revisited: The Cascade Construction and its Concrete Security. In *37th IEEE Symposium on Foundations of Computer Science*, pages 514–523, 1996.

33. M. Bellare, R. Canetti and H. Krawczyk. Keying Hash Functions for Message Authentication. In *Crypto96*, Springer Lecture Notes in Computer Science (Vol. 1109), pages 1–15.

34. M. Bellare, R. Canetti and H. Krawczyk. Modular Approach to the Design and Analysis of Authentication and Key Exchange Protocols. In *30th ACM Symposium on the Theory of Computing*, pages 419–428, 1998.

35. M. Bellare, A. Desai, D. Pointcheval and P. Rogaway. Relations among notions of security for public-key encryption schemes. In *Crypto98*,

36. M. Bellare and O. Goldreich. On Defining Proofs of Knowledge. In *Crypto92*, Springer-Verlag Lecture Notes in Computer Science (Vol. 740), pages 390–420.

37. M. Bellare, O. Goldreich, and S. Goldwasser. Randomness in Interactive Proofs. *Computational Complexity*, Vol. 4, No. 4, pages 319–354, 1993.

38. M. Bellare, O. Goldreich and S. Goldwasser. Incremental Cryptography: the Case of Hashing and Signing. In *Crypto94*, Springer-Verlag Lecture Notes in Computer Science (Vol. 839), pages 216–233, 1994.

39. M. Bellare, O. Goldreich and S. Goldwasser. Incremental Cryptography and Application to Virus Protection. In *27th ACM Symposium on the Theory of Computing*, pages 45–56, 1995.

40. M. Bellare, O. Goldreich and M. Sudan. Free Bits, PCPs and Non-Approximability – Towards Tight Results. *SIAM Journal on Computing*, Vol. 27, No. 3, pages 804–915, 1998.

41. M. Bellare and S. Goldwasser. The Complexity of Decision versus Search. *SIAM Journal on Computing*, Vol. 23, pages 97–119, 1994.

42. M. Bellare, S. Goldwasser, C. Lund and A. Russell. Efficient probabilistically checkable proofs and applications to approximation. In *25th ACM Symposium on the Theory of Computing*, pages 294–304, 1993.

43. M. Bellare, S. Goldwasser and D. Micciancio. "Pseudo-random" Number Generation within Cryptographic Algorithms: the DSS Case. In *Crypto97*, Springer Lecture Notes in Computer Science (Vol. 1294), pages 277–291.

44. M. Bellare, R. Guerin and P. Rogaway. XOR MACs: New Methods for Message Authentication using Finite Pseudorandom Functions. In *Crypto95*, Springer-Verlag Lecture Notes in Computer Science (Vol. 963), pages 15–28.

45. M. Bellare, S. Halevi, A. Sahai and S. Vadhan. Trapdoor Functions and Public-Key Cryptosystems. In *Crypto98*,

46. M. Bellare, R. Impagliazzo and M. Naor. Does Parallel Repetition Lower the Error in Computationally Sound Protocols? In *38th IEEE Symposium on Foundations of Computer Science*, pages 374–383, 1997.

47. M. Bellare, J. Kilian and P. Rogaway. The Security of Cipher Block Chaining. In *Crypto94*, Springer-Verlag Lecture Notes in Computer Science (Vol. 839), pages 341–358.

48. M. Bellare and S. Micali. How to Sign Given Any Trapdoor Function. *Journal of the ACM*, Vol. 39, pages 214–233, 1992.

49. M. Bellare and P. Rogaway. Random Oracles are Practical: a Paradigm for Designing Efficient Protocols. In *1st Conf. on Computer and Communications Security*, ACM, pages 62–73, 1993.

50. M. Bellare and P. Rogaway. Entity Authentication and Key Distribution. In *Crypto93*, Springer-Verlag Lecture Notes in Computer Science (Vol. 773), pages 232–249, 1994.

51. M. Bellare and P. Rogaway. Provably Secure Session Key Distribution: The Three Party Case. In *27th ACM Symposium on the Theory of Computing*, pages 57–66, 1995.

52. M. Bellare and P. Rogaway. The Exact Security of Digital Signatures: How to Sign with RSA and Rabin. In *EuroCrypt96*, Springer Lecture Notes in Computer Science (Vol. 1070), pages 399–416.

53. M. Bellare and J. Rompel. Randomness-efficient oblivious sampling. In *35th IEEE Symposium on Foundations of Computer Science*, pages 276–287, 1994.

54. M. Bellare and M. Sudan. Improved non-approximability results. In *26th ACM Symposium on the Theory of Computing*, pages 184–193, 1994.

55. C.H. Bennett, G. Brassard and J.M. Robert. Privacy Amplification by Public Discussion. *SIAM Journal on Computing*, Vol. 17, pages 210–229, 1988. Preliminary version in *Crypto85*, Springer-Verlag Lecture Notes in Computer Science (Vol. 218), pages 468–476 (titled "How to Reduce your Enemy's Information").

56. M. Ben-Or. Another advantage of free choice: Completely Asynchronous Byzantine Agreement. In *2nd ACM Symposium on Principles of Distributed Computing*, pages 27–30, 1983.

57. M. Ben-Or, O. Goldreich, S. Goldwasser, J. Håstad, J. Kilian, S. Micali and P. Rogaway. Everything Provable is Probable in Zero-Knowledge. In *Crypto88*, Springer-Verlag Lecture Notes in Computer Science (Vol. 403), pages 37–56, 1990

58. M. Ben-Or, S. Goldwasser, J. Kilian and A. Wigderson. Multi-Prover Interactive Proofs: How to Remove Intractability. In *20th ACM Symposium on the Theory of Computing*, pages 113–131, 1988.

59. M. Ben-Or, S. Goldwasser and A. Wigderson. Completeness Theorems for Non-Cryptographic Fault-Tolerant Distributed Computation. In *20th ACM Symposium on the Theory of Computing*, pages 1–10, 1988.

60. G.R. Blakley. Safeguarding Cryptographic Keys. In *Proc. of National Computer Conf.*, Vol. 48, AFIPS Press, pages 313–317, 1979.

61. M. Blum. How to Exchange Secret Keys. *ACM Trans. Comput. Sys.*, Vol. 1, pages 175–193, 1983.

62. M. Blum. Coin Flipping by Phone. *IEEE Spring COMPCOM*, pages 133–137, February 1982. See also *SIGACT News*, Vol. 15, No. 1, 1983.

63. L. Blum, M. Blum and M. Shub. A Simple Secure Unpredictable Pseudo-Random Number Generator. *SIAM Journal on Computing*, Vol. 15, 1986, pages 364–383.

64. M. Blum, A. De Santis, S. Micali, and G. Persiano. Non-Interactive Zero-Knowledge Proof Systems. *SIAM Journal on Computing*, Vol. 20, No. 6, pages 1084–1118, 1991. (Considered the journal version of [66].)

65. M. Blum, W. Evans, P. Gemmell, S. Kannan and M. Naor. Checking the correctness of memories. In *32nd IEEE Symposium on Foundations of Computer Science*, pages 90-99, 1991.

66. M. Blum, P. Feldman and S. Micali. Non-Interactive Zero-Knowledge and its Applications. In *20th ACM Symposium on the Theory of Computing*, pages 103–112, 1988. See [64].

67. M. Blum and O. Goldreich. Towards a Computational Theory of Statistical Tests. In *33rd IEEE Symposium on Foundations of Computer Science*, pages 406–416, 1992.

68. M. Blum and S. Goldwasser. An Efficient Probabilistic Public-Key Encryption Scheme which hides all partial information. In *Crypto84*, Lecture Notes in Computer Science (Vol. 196) Springer-Verlag, pages 289–302.

69. M. Blum, M. Luby and R. Rubinfeld. Self-Testing/Correcting with Applications to Numerical Problems. *Journal of Computer and System Science*, Vol. 47, No. 3, pages 549–595, 1993.

70. M. Blum and S. Kannan. Designing Programs that Check their Work. In *21st ACM Symposium on the Theory of Computing*, pages 86–97, 1989.

71. M. Blum and S. Micali. How to Generate Cryptographically Strong Sequences of Pseudo-Random Bits. *SIAM Journal on Computing*, Vol. 13, pages 850–864, 1984. Preliminary version in *23rd IEEE Symposium on Foundations of Computer Science*, 1982.

72. D. Boneh, R. DeMillo and R. Lipton. On the Importance of Checking Cryptographic Protocols for Faults. In *EuroCrypt97*, Springer Lecture Notes in Computer Science (Vol. 1233), pages 37–51, 1997.

73. R. Boppana, J. Håstad, and S. Zachos. Does Co-NP Have Short Interactive Proofs? *Information Processing Letters*, 25, May 1987, pp. 127-132.

74. J.B. Boyar. Inferring Sequences Produced by Pseudo-Random Number Generators. *Journal of the ACM*, Vol. 36, pages 129–141, 1989.

75. G. Brassard. A Note on the Complexity of Cryptography. *IEEE Trans. on Inform. Th.*, Vol. 25, pages 232–233, 1979.

76. G. Brassard. Quantum Information Processing: The Good, the Bad and the Ugly. In *Crypto97*, Springer Lecture Notes in Computer Science (Vol. 1294), pages 337–341.

77. G. Brassard, D. Chaum and C. Crépeau. Minimum Disclosure Proofs of Knowledge. *Journal of Computer and System Science*, Vol. 37, No. 2, pages 156–189, 1988. Preliminary version by Brassard and Crépeau in *27th IEEE Symposium on Foundations of Computer Science*, 1986.

78. G. Brassard and C. Crépeau. Zero-Knowledge Simulation of Boolean Circuits. In *Crypto86*, Springer-Verlag Lecture Notes in Computer Science (Vol. 263), pages 223–233, 1987.

79. G. Brassard, C. Crépeau and M. Yung. Constant-Round Perfect Zero-Knowledge Computationally Convincing Protocols. *Theoretical Computer Science*, Vol. 84, pages 23–52, 1991.

80. C. Cachin and U. Maurer. Unconditional security against memory-bounded adversaries. In *Crypto97*, Springer Lecture Notes in Computer Science (Vol. 1294), pages 292–306.

81. R. Canetti. *Studies in Secure Multi-Party Computation and Applications*. Ph.D. Thesis, Department of Computer Science and Applied Mathematics, Weizmann Institute of Science, Rehovot, Israel, June 1995. Available from http://theory.lcs.mit.edu/~tcryptol/BOOKS/ran-phd.html.

82. R. Canetti. Towards Realizing Random Oracles: Hash Functions that Hide All Partial Information. In *Crypto97*, Springer Lecture Notes in Computer Science (Vol. 1294), pages 455–469.

83. R. Canetti. Security and Composition of Multi-party Cryptographic Protocols. Record 98-18 of the *Theory of Cryptography Library*, URL http://theory.lcs.mit.edu/~tcryptol. June 1998.

84. R. Canetti, C. Dwork, M. Naor and R. Ostrovsky. Deniable Encryption. In *Crypto97*, Springer Lecture Notes in Computer Science (Vol. 1294), pages 90–104.

85. R. Canetti, G. Even and O. Goldreich. Lower Bounds for Sampling Algorithms for Estimating the Average. *Information Processing Letters*, Vol. 53, pages 17–25, 1995.

86. R. Canetti, U. Feige, O. Goldreich and M. Naor. Adaptively Secure Multiparty Computation. In *28th ACM Symposium on the Theory of Computing*, pages 639–648, 1996.

87. R. Canetti and R. Gennaro. Incoercible Multiparty Computation. In *37th IEEE Symposium on Foundations of Computer Science*, pages 504–513, 1996.

88. R. Canetti, O. Goldreich and S. Halevi. The Random Oracle Methodology, Revisited. In *30th ACM Symposium on the Theory of Computing*, pages 209–218, 1998.

89. R. Canetti, D. Micciancio and O. Reingold. Using one-way functions to construct Hash Functions that Hide All Partial Information. In *30th ACM Symposium on the Theory of Computing*, pages 131–140, 1998.

90. R. Canetti, S. Halevi and A. Herzberg. How to Maintain Authenticated Communication in the Presence of Break-Ins. In *16th ACM Symposium on Principles of Distributed Computing*, pages 15–24, 1997.

91. R. Canetti and A. Herzberg. Maintaining Security in the Presence of Transient Faults. In *Crypto94*, Springer-Verlag Lecture Notes in Computer Science (Vol. 839), pages 425–439.

92. L. Carter and M. Wegman. Universal Hash Functions. *Journal of Computer and System Science*, Vol. 18, 1979, pages 143–154.

93. G.J. Chaitin. On the Length of Programs for Computing Finite Binary Sequences. *Journal of the ACM*, Vol. 13, pages 547–570, 1966.

94. A.K. Chandra, D.C. Kozen and L.J. Stockmeyer. Alternation. *Journal of the ACM*, Vol. 28, pages 114–133, 1981.

95. S. Chari, P. Rohatgi and A. Srinivasan. Improved Algorithms via Approximation of Probability Distributions. In *26th ACM Symposium on the Theory of Computing*, pages 584–592, 1994.

96. D. Chaum. Blind Signatures for Untraceable Payments. In *Crypto82*, Plenum Press, pages 199–203, 1983.

97. D. Chaum, C. Crépeau and I. Damgård. Multi-party unconditionally Secure Protocols. In *20th ACM Symposium on the Theory of Computing*, pages 11–19, 1988.

98. D. Chaum, A. Fiat and M. Naor. Untraceable Electronic Cash. In *Crypto88*, Springer-Verlag Lecture Notes in Computer Science (Vol. 403), pages 319–327.

99. R. Chang, B. Chor, O. Goldreich, J. Hartmanis, J. Håstad, D. Ranjan, and P. Rohatgi. The Random Oracle Hypothesis is False. *Journal of Computer and System Science*, Vol. 49, No. 1, pages 24–39, 1994.

100. B. Chor and C. Dwork. Randomization in Byznatine Agreement. *Advances in Computing Research: A Research Annual*, Vol. 5 (Randomness and Computation, S. Micali, ed.), pages 443–497, 1989.

101. B. Chor, J. Friedmann, O. Goldreich, J. Håstad, S. Rudich and R. Smolensky. The bit extraction problem and t-resilient functions. In *26th IEEE Symposium on Foundations of Computer Science*, pages 396–407, 1985.

102. B. Chor and N. Gilboa. Computationally Private Information Retrieval. In *29th ACM Symposium on the Theory of Computing*, pages 304–313, 1997.

103. B. Chor and O. Goldreich. On the Power of Two–Point Based Sampling. *Jour. of Complexity*, Vol 5, 1989, pages 96–106. Preliminary version dates 1985.

104. B. Chor and O. Goldreich. Unbiased Bits from Sources of Weak Randomness and Probabilistic Communication Complexity. *SIAM Journal on Computing*, Vol. 17, No. 2, pages 230–261, 1988.

105. B. Chor, O. Goldreich, E. Kushilevitz and M. Sudan, Private Information Retrieval. In *36th IEEE Symposium on Foundations of Computer Science*, pages 41–50, 1995.

106. B. Chor, S. Goldwasser, S. Micali and B. Awerbuch. Verifiable Secret Sharing and Achieving Simultaneity in the Presence of Faults. In *26th IEEE Symposium on Foundations of Computer Science*, pages 383–395, 1985.

107. R. Cleve. Limits on the Security of Coin Flips when Half the Processors are Faulty. In *18th ACM Symposium on the Theory of Computing*, pages 364–369, 1986.

108. A. Cohen and A. Wigderson. Dispensers, Deterministic Amplification, and Weak Random Sources. *30th IEEE Symposium on Foundations of Computer Science*, 1989, pages 14–19.

109. T.M. Cover and G.A. Thomas. *Elements of Information Theory*. John Wiley & Sons, Inc., New-York, 1991.

110. R. Cramer and I. Damgård. New Generation of Secure and Practical RSA-based Signatures. In *Crypto96*, Springer Lecture Notes in Computer Science (Vol. 1109), pages 173–185.

111. R. Cramer and I. Damgård. Linear Zero-Knowledge – A Note on Efficient Zero-Knowledge Proofs and Arguments. In *29th ACM Symposium on the Theory of Computing*, pages 436–445, 1997.

112. R. Cramer and I. Damgård. Zero-Knowledge Proofs for Finite Field Arithmetic; or: Can Zero-Knowledge be for Free? In *Crypto98*,

113. R. Cramer, I. Damgård, and T. Pedersen. Efficient and provable security amplifications. In *Proc. of 4th Cambridge Security Protocols Workshop*, Springer, Lecture Notes in Computer Science (Vol. 1189), pages 101–109.

114. C. Crépeau. Efficient Cryptographic Protocols Based on Noisy Channels. In *EuroCrypt97*, Springer, Lecture Notes in Computer Science (Vol. 1233), pages 306–317.

115. I. Damgård. Collision Free Hash Functions and Public Key Signature Schemes. In *EuroCrypt87*, Springer-Verlag, Lecture Notes in Computer Science (Vol. 304), pages 203–216.

116. I. Damgård. A Design Principle for Hash Functions. In *Crypto89*, Springer-Verlag Lecture Notes in Computer Science (Vol. 435), pages 416–427.

117. I. Damgård, O. Goldreich, T. Okamoto and A. Wigderson. Honest Verifier vs Dishonest Verifier in Public Coin Zero-Knowledge Proofs. In *Crypto95*, Springer-Verlag Lecture Notes in Computer Science (Vol. 963), pages 325–338, 1995.

118. A. De-Santis, Y. Desmedt, Y. Frankel and M. Yung. How to Share a Function Securely. In *26th ACM Symposium on the Theory of Computing*, pages 522–533, 1994.

119. Y. Desmedt. Society and group oriented cryptography: A new concept. In *Crypto87*, Springer-Verlag, Lecture Notes in Computer Science (Vol. 293), pages 120–127.

120. Y. Desmedt and Y. Frankel. Threshold Cryptosystems. In *Crypto89*, Springer-Verlag Lecture Notes in Computer Science (Vol. 435), pages 307–315.

121. W. Diffie, and M.E. Hellman. New Directions in Cryptography. *IEEE Trans. on Info. Theory*, IT-22 (Nov. 1976), pages 644–654.

122. D. Dolev, C. Dwork, and M. Naor. Non-Malleable Cryptography. In *23rd ACM Symposium on the Theory of Computing*, pages 542–552, 1991. Full version available from authors.

123. D. Dolev, M.J. Fischer, R. Fowler, N.A. Lynch and H.R. Strong. An efficient algorithm for Byzantine Agreement without authentication. *Information and Control*, Vol. 52(3), pages 257–274, March 1982.

124. D. Dolev and H.R. Strong. Authenticated Algorithms for Byzantine Agreement. *SIAM Journal on Computing*, Vol. 12, pages 656–666, 1983.

125. D. Dolev and A.C. Yao. On the Security of Public-Key Protocols. *IEEE Trans. on Inform. Theory*, Vol. 30, No. 2, pages 198–208, 1983.

126. C. Dwork, U. Feige, J. Kilian, M. Naor and S. Safra. Low Communication Perfect Zero Knowledge Two Provers Proof Systems. In *Crypto92*, Springer Verlag, LNCS Vol. 740, pages 215–227, 1992.

127. C. Dwork, and M. Naor. An Efficient Existentially Unforgeable Signature Scheme and its Application. To appear in *Journal of Cryptology*. Preliminary version in *Crypto94*.

128. G. Even, O. Goldreich, M. Luby, N. Nisan, and B. Veličković. Approximations of General Independent Distributions. In *24th ACM Symposium on the Theory of Computing*, pages 10–16, 1992. Revised version available from http://theory.lcs.mit.edu/~oded/papers.html,

129. S. Even and O. Goldreich. On the Security of Multi-party Ping-Pong Proto-
     cols. In *24th IEEE Symposium on Foundations of Computer Science*, pages
     34–39, 1983.
130. S. Even, O. Goldreich, and A. Lempel. A Randomized Protocol for Signing
     Contracts. *Communications of the ACM*, Vol. 28, No. 6, 1985, pages 637–647.
131. S. Even, O. Goldreich and S. Micali. On-line/Off-line Digital signatures. *Jour-
     nal of Cryptology*, Vol. 9, 1996, pages 35–67.
132. S. Even, A.L. Selman, and Y. Yacobi. The Complexity of Promise Problems
     with Applications to Public-Key Cryptography. *Inform. and Control*, Vol. 61,
     pages 159–173, 1984.
133. S. Even and Y. Yacobi. Cryptography and NP-Completeness. In proceedings
     of *7th ICALP*, Springer-Verlag Lecture Notes in Computer Science (Vol. 85),
     pages 195–207, 1980. See [132].
134. U. Feige. A Threshold of ln n for Approximating Set Cover. In *28th ACM
     Symposium on the Theory of Computing*, pages 314–318, 1996.
135. U. Feige. On the success probability of the two provers in One-Round Proof
     Systems. In *Proc. 6th IEEE Symp. on Structure in Complexity Theory*, pages
     116–123, 1991.
136. U. Feige. Error reduction by parallel repetition – the state of the art. Tech-
     nical report CS95-32, Computer Science Department, Weizmann Institute of
     Science, Rehovot, ISREAL, 1995.
137. U. Feige, A. Fiat and A. Shamir. Zero-Knowledge Proofs of Identity. *Journal
     of Cryptology*, Vol. 1, 1988, pages 77–94.
138. U. Feige, S. Goldwasser, L. Lovász and S. Safra. On the Complexity of Ap-
     proximating the Maximum Size of a Clique. Unpublished manuscript, 1990.
139. U. Feige, S. Goldwasser, L. Lovász, S. Safra, and M. Szegedy. Approximating
     Clique is almost NP-complete. In *32nd IEEE Symposium on Foundations of
     Computer Science*, pages 2–12, 1991.
140. U. Feige and J. Kilian. Two prover protocols – Low error at affordable rates.
     In *26th ACM Symposium on the Theory of Computing*, pages 172–183, 1994.
141. U. Feige and J. Kilian. Zero knowledge and the chromatic number. In *11th
     IEEE Conference on Computational Complexity*, pages 278–287, 1996.
142. U. Feige and J. Kilian. Making games short (extended abstract). In *29th
     ACM Symposium on the Theory of Computing*, pages 506–516, 1997.
143. U. Feige, D. Lapidot, and A. Shamir. Multiple Non-Interactive Zero-
     Knowledge Proofs Based on a Single Random String. In *31th IEEE Sym-
     posium on Foundations of Computer Science*, pages 308–317, 1990. To appear
     in *SIAM Journal on Computing*.
144. U. Feige and A. Shamir. Zero-Knowledge Proofs of Knowledge in Two Rounds.
     In *Crypto89*, Springer-Verlag Lecture Notes in Computer Science (Vol. 435),
     pages 526–544.
145. U. Feige and A. Shamir. Witness Indistinguishability and Witness Hiding
     Protocols. In *22nd ACM Symposium on the Theory of Computing*, pages 416–
     426, 1990.
146. U. Feige, A. Shamir and M. Tennenholtz. The noisy oracle problem. In
     Crypto88, Springer-Verlag Lecture Notes in Computer Science (Vol. 403), pa-
     ges 284–296.
147. P. Feldman. A Practical Scheme for Non-interactive Verifiable Secret Sharing.
     In *28th IEEE Symposium on Foundations of Computer Science*, pages 427–
     437, 1987.
148. P. Feldman and S. Micali. An optimal probabilistic protocol for synchronous
     Byzantine Agreement. *SICOMP*, Vol. 26, pages 873–933, 1997.
149. A. Fiat. Batch RSA. *Journal of Cryptology*, Vol. 10, 1997, pages 75–88.

150. A. Fiat and A. Shamir. How to Prove Yourself: Practical Solution to Identification and Signature Problems. In *Crypto86*, Springer-Verlag Lecture Notes
in Computer Science (Vol. 263), pages 186–189, 1987.

151. J.B. Fischer and J. Stern. An Efficient Pseudorandom Generator Provably
as Secure as Syndrome Decoding. In *EuroCrypt96*, Springer Lecture Notes in
Computer Science (Vol. 1070), pages 245–255.

152. R. Fischlin and C.P. Schnorr. Stronger Security Proofs for RSA and Rabin Bits. In *EuroCrypt97*, Springer Lecture Notes in Computer Science
(Vol. 1233), pages 267–279, 1997.

153. L. Fortnow, The Complexity of Perfect Zero-Knowledge. In *19th ACM Symposium on the Theory of Computing*, pages 204–209, 1987.

154. L. Fortnow, J. Rompel and M. Sipser. On the power of multi-prover interactive
protocols. In *Proc. 3rd IEEE Symp. on Structure in Complexity Theory*, pages
156–161, 1988.

155. L. Fortnow, J. Rompel and M. Sipser. Errata for "On the power of multi-prover
interactive protocols." In *Proc. 5th IEEE Symp. on Structure in Complexity
Theory*, pages 318–319, 1990.

156. M. Franklin and M. Yung. Secure and Efficient Off-Line Digital Money. In
20th ICALP, Springer-Verlag Lecture Notes in Computer Science (Vol. 700),
pages 265–276.

157. A.M. Frieze, J. Håstad, R. Kannan, J.C. Lagarias, and A. Shamir. Recon-
structing Truncated Integer Variables Satisfying Linear Congruences. *SIAM
Journal on Computing*, Vol. 17, pages 262–280, 1988.

158. M. Fürer, O. Goldreich, Y. Mansour, M. Sipser, and S. Zachos. On Completeness and Soundness in Interactive Proof Systems. *Advances in Computing
Research: a research annual*, Vol. 5 (Randomness and Computation, S. Micali,
ed.), pages 429–442, 1989.

159. O. Gaber and Z. Galil. Explicit Constructions of Linear Size Superconcentrators. *Journal of Computer and System Science*, Vol. 22, pages 407–420,
1981.

160. P.S. Gemmell. An Introduction to Threshold Cryptography. In *CryptoBytes*,
RSA Lab., Vol. 2, No. 3, 1997.

161. P. Gemmell, R. Lipton, R. Rubinfeld, M. Sudan, and A. Wigderson. Self-
Testing/Correcting for Polynomials and for Approximate Functions. In *23th
ACM Symposium on the Theory of Computing*, pages 32–42, 1991.

162. R. Gennaro, S. Jarecki, H. Krawczyk, and T. Rabin. Robust Threshold DSS
Signatures. In *EuroCrypt96*, Springer-Verlag, Lecture Notes in Computer
Science (Vol. 1070), pages 354–371.

163. M. Goemans and D. Williamson. New 3/4-approximation algorithms for the
maximum satisfiablity problem. *SIAM Journal on Discrete Mathematics*,
Vol. 7, No. 4, pages 656–666, 1994.

164. M. Goemans and D. Williamson. Improved approximation algorithms for maximum cut and satisfiability problems using semidefinite programming. *Journal of the ACM*, Vol. 42, No. 6, 1995, pages 1115–1145.

165. O. Goldreich. Two Remarks Concerning the GMR Signature Scheme. In
Crypto86, Springer-Verlag Lecture Notes in Computer Science (Vol. 263), pages 104–110, 1987.

166. O. Goldreich. A Note on Computational Indistinguishability. *Information
Processing Letters*, Vol. 34, pages 277–281, May 1990.

167. O. Goldreich. *Lecture Notes on Encryption, Signatures and Cryptographic
Protocol.* Spring 1989.
Available from http://theory.lcs.mit.edu/~oded/ln89.html.

168. O. Goldreich. A Uniform Complexity Treatment of Encryption and Zero-Knowledge. *Journal of Cryptology*, Vol. 6, No. 1, pages 21–53, 1993.
169. O. Goldreich. Three XOR-Lemmas – An Exposition. *ECCC*, TR95-056, 1995. Available from `http://www.eccc.uni-trier.de/eccc/`.
170. O. Goldreich. *Foundation of Cryptography – Fragments of a Book.* February 1995. Revised version, January 1998. Both versions are available from `http://theory.lcs.mit.edu/~oded/frag.html`.
171. O. Goldreich. A Sample of Samplers – A Computational Perspective on Sampling. *ECCC*, TR97-020, May 1997.
172. O. Goldreich. Notes on Levin's Theory of Average-Case Complexity. *ECCC*, TR97-058, Dec. 1997.
173. O. Goldreich. *Secure Multi-Party Computation.* In preparation, 1998. Working draft available from `http://theory.lcs.mit.edu/~oded/gmw.html`.
174. O. Goldreich, S. Goldwasser, and S. Micali. How to Construct Random Functions. *Journal of the ACM*, Vol. 33, No. 4, pages 792–807, 1986.
175. O. Goldreich, S. Goldwasser, and S. Micali. On the Cryptographic Applications of Random Functions. In *Crypto84*, Springer-Verlag Lecture Notes in Computer Science (Vol. 263), pages 276–288, 1985.
176. O. Goldreich and J. Håstad. On the Message Complexity of Interactive Proof Systems. To appear in *Information Processing Letters*. Available as TR96-018 of *ECCC*, `http://www.eccc.uni-trier.de/eccc/`, 1996.
177. O. Goldreich, R. Impagliazzo, L.A. Levin, R. Venkatesan, and D. Zuckerman. Security Preserving Amplification of Hardness. In *31st IEEE Symposium on Foundations of Computer Science*, pages 318–326, 1990.
178. O. Goldreich and A. Kahan. How to Construct Constant-Round Zero-Knowledge Proof Systems for NP. *Journal of Cryptology*, Vol. 9, No. 2, pages 167–189, 1996. Preliminary versions date to 1988.
179. O. Goldreich and H. Krawczyk. On the Composition of Zero-Knowledge Proof Systems. *SIAM Journal on Computing*, Vol. 25, No. 1, February 1996, pages 169–192. Preliminary version in *17th ICALP*, 1990.
180. O. Goldreich, and H. Krawczyk, On Sparse Pseudorandom Ensembles. *Random Structures and Algorithms*, Vol. 3, No. 2, (1992), pages 163–174.
181. O. Goldreich, H. Krawcyzk and M. Luby. On the Existence of Pseudorandom Generators. *SIAM Journal on Computing*, Vol. 22-6, pages 1163–1175, 1993.
182. O. Goldreich and L.A. Levin. Hard-core Predicates for any One-Way Function. In *21st ACM Symposium on the Theory of Computing*, pages 25–32, 1989.
183. O. Goldreich and B. Meyer. Computational Indistinguishability – Algorithms vs. Circuits. *Theoretical Computer Science*, Vol. 191, pages 215–218, 1998. Preliminary version by Meyer in *Structure in Complexity Theory*, 1994.
184. O. Goldreich and S. Micali. Increasing the Expansion of Pseudorandom Generators. Manuscript, 1984. Available from `http://theory.lcs.mit.edu/~oded/papers.html`
185. O. Goldreich, S. Micali and A. Wigderson. Proofs that Yield Nothing but their Validity or All Languages in NP Have Zero-Knowledge Proof Systems. *Journal of the ACM*, Vol. 38, No. 1, pages 691–729, 1991. Preliminary version in *27th IEEE Symposium on Foundations of Computer Science*, 1986.
186. O. Goldreich, S. Micali and A. Wigderson. How to Play any Mental Game – A Completeness Theorem for Protocols with Honest Majority. In *19th ACM Symposium on the Theory of Computing*, pages 218–229, 1987.
187. O. Goldreich and Y. Oren. Definitions and Properties of Zero-Knowledge Proof Systems. *Journal of Cryptology*, Vol. 7, No. 1, pages 1–32, 1994.
188. O. Goldreich and R. Ostrovsky. Software Protection and Simulation on Oblivious RAMs. *Journal of the ACM*, Vol. 43, 1996, pages 431–473.

189. O. Goldreich and E. Petrank. Quantifying Knowledge Complexity. In *32nd IEEE Symposium on Foundations of Computer Science*, pp. 59–68, 1991.

190. O. Goldreich, R. Rubinfeld and M. Sudan. Learning polynomials with queries: the highly noisy case. In *36th IEEE Symposium on Foundations of Computer Science*, pages 294–303, 1995.

191. O. Goldreich and S. Safra. A Combinatorial Consistency Lemma with application to the PCP Theorem. In the proceedings of *Random97*, Springer Lecture Notes in Computer Science (Vol. 1269), pages 67–84. See also *ECCC*, TR96-047, 1996.

192. O. Goldreich, A. Sahai, and S. Vadhan. Honest-Verifier Statistical Zero-Knowledge equals general Statistical Zero-Knowledge. In *30th ACM Symposium on the Theory of Computing*, pages 399–408, 1998.

193. O. Goldreich and M. Sudan. Computational Indistinguishability: k versus $2k$ samples. In *13th IEEE Conference on Computational Complexity*, to appear, 1998.

194. O. Goldreich and A. Wigderson. Tiny Families of Functions with Random Properties: A Quality–Size Trade–off for Hashing. *Journal of Random structures and Algorithms*, Vol. 11, Nr. 4, December 1997, pages 315–343.

195. O. Goldreich and D. Zuckerman. Another proof that BPP subseteq PH (and more). *ECCC*, TR97-045, 1997.

196. S. Goldwasser. Fault Tolerant Multi Party Computations: Past and Present. In *16th ACM Symposium on Principles of Distributed Computing*, pages 1–6, 1997.

197. S. Goldwasser and L.A. Levin. Fair Computation of General Functions in Presence of Immoral Majority. In *Crypto90*, Springer-Verlag Lecture Notes in Computer Science (Vol. 537), pages 77–93.

198. S. Goldwasser and S. Micali. Probabilistic Encryption. *Journal of Computer and System Science*, Vol. 28, No. 2, pages 270–299, 1984. Preliminary version in *14th ACM Symposium on the Theory of Computing*, 1982.

199. S. Goldwasser, S. Micali and C. Rackoff. The Knowledge Complexity of Interactive Proof Systems. *SIAM Journal on Computing*, Vol. 18, pages 186–208, 1989. Preliminary version in *17th ACM Symposium on the Theory of Computing*, 1985. Earlier versions date to 1982.

200. S. Goldwasser, S. Micali, and R.L. Rivest. A Digital Signature Scheme Secure Against Adaptive Chosen-Message Attacks. *SIAM Journal on Computing*, April 1988, pages 281–308.

201. S. Goldwasser, S. Micali and P. Tong. Why and How to Establish a Private Code in a Public Network. In *23rd IEEE Symposium on Foundations of Computer Science*, 1982, pages 134–144.

202. S. Goldwasser, S. Micali and A.C. Yao. Strong Signature Schemes. In *15th ACM Symposium on the Theory of Computing*, pages 431–439, 1983.

203. S. Goldwasser and M. Sipser. Private Coins versus Public Coins in Interactive Proof Systems. *Advances in Computing Research: a research annual*, Vol. 5 (Randomness and Computation, S. Micali, ed.), pages 73–90, 1989. Extended abstract in *18th ACM Symposium on the Theory of Computing*, pages 59–68, 1986.

204. S. W. Golomb. *Shift Register Sequences*. Holden-Day, 1967. (Aegean Park Press, Revised edition, 1982.)

205. V. Guruswami, D. Lewin, M. Sudan and L. Trevisan. A tight characterization of NP with 3 query PCPs. To appear in *39th IEEE Symposium on Foundations of Computer Science*, 1998.

206. S. Hada and T. Tanaka. On the Existence of 3-Round Zero-Knowledge Protocols. In *Crypto98*,

207. J. Håstad. Almost optimal lower bounds for small depth circuits. *Advances in Computing Research: a research annual*, Vol. 5 (Randomness and Computation, S. Micali, ed.), pages 143–170, 1989. Extended abstract in *18th ACM Symposium on the Theory of Computing*, pages 6–20, 1986.

208. J. Håstad. Pseudo-Random Generators under Uniform Assumptions. In *22nd ACM Symposium on the Theory of Computing*, pages 395–404, 1990.

209. J. Håstad. Clique is hard to approximate within $n^{1-\epsilon}$. To appear in *ACTA Mathematica*. Preliminary versions in *28th ACM Symposium on the Theory of Computing* (1996) and *37th IEEE Symposium on Foundations of Computer Science* (1996).

210. J. Håstad. Getting optimal in-approximability results. In *29th ACM Symposium on the Theory of Computing*, pages 1–10, 1997.

211. J. Håstad, R. Impagliazzo, L.A. Levin and M. Luby. Construction of Pseudo-random Generator from any One-Way Function. To appear in *SIAM Journal on Computing*. Combines the results of [217] and [208].

212. J. Håstad, S. Phillips and S. Safra. A Well Characterized Approximation Problem. *Information Processing Letters*, Vol. 47:6, pages 301–305. 1993.

213. J. Håstad, A. Schrift and A. Shamir. The Discrete Logarithm Modulo a Composite Hides $O(n)$ Bits. *Journal of Computer and System Science*, Vol. 47, pages 376–404, 1993.

214. A. Herzberg, M. Jakobsson, S. Jarecki, H. Krawczyk and M. Yung. Proactive public key and signature systems. In *1997 ACM Conference on Computers and Communication Security*, pages 100–110, 1997.

215. A. Herzberg, S. Jarecki, H. Krawczyk and M. Yung. Proactive Secret Sharing, or How to Cope with Perpetual Leakage. In *Crypto95*, Springer-Verlag Lecture Notes in Computer Science (Vol. 963), pages 339–352.

216. R. Impagliazzo. Hard-core Distributions for Somewhat Hard Problems. In *36th IEEE Symposium on Foundations of Computer Science*, pages 538–545, 1995.

217. R. Impagliazzo, L.A. Levin and M. Luby. Pseudorandom Generation from One-Way Functions. In *21st ACM Symposium on the Theory of Computing*, pages 12–24, 1989.

218. R. Impagliazzo and M. Luby. One-Way Functions are Essential for Complexity Based Cryptography. In *30th IEEE Symposium on Foundations of Computer Science*, pages 230–235, 1989.

219. R. Impagliazzo and M. Naor. Efficient Cryptographic Schemes Provable as Secure as Subset Sum. *Journal of Cryptology*, Vol. 9, 1996, pages 199–216.

220. R. Impagliazzo and S. Rudich. Limits on the Provable Consequences of One-Way Permutations. In *21st ACM Symposium on the Theory of Computing*, pages 44–61, 1989.

221. R. Impagliazzo and A. Wigderson. P=BPP if E requires exponential circuits: Derandomizing the XOR Lemma. In *29th ACM Symposium on the Theory of Computing*, pages 220–229, 1997.

222. R. Impagliazzo and M. Yung. Direct Zero-Knowledge Computations. In *Crypto87*, Springer-Verlag Lecture Notes in Computer Science (Vol. 293), pages 40–51, 1987.

223. R. Impagliazzo and D. Zuckerman. How to Recycle Random Bits. In *30th IEEE Symposium on Foundations of Computer Science*, 1989, pages 248–253.

224. A. Juels, M. Luby and R. Ostrovsky. Security of Blind Digital Signatures. In *Crypto97*, Springer Lecture Notes in Computer Science (Vol. 1294), pages 150–164.

225. J. Justesen. A class of constructive asymptotically good alegbraic codes. *IEEE Trans. Inform. Theory*, Vol. 18, pages 652–656, 1972.

226. N. Kahale, Eigenvalues and Expansion of Regular Graphs. *Journal of the ACM*, 42(5):1091–1106, September 1995.

227. D.R. Karger. Global Min-cuts in RNC, and Other Ramifications of a Simple Min-Cut Algorithm. In *4th SODA*, pages 21–30, 1993.

228. H. Karloff and U. Zwick. A 7/8-approximation algorithm for MAX 3SAT? In *38th IEEE Symposium on Foundations of Computer Science*, 1997, pages 406–415.

229. R.M. Karp and M. Luby. Monte-Carlo algorithms for enumeration and reliability problems. In *24th IEEE Symposium on Foundations of Computer Science*, pages 56-64, 1983. See [230].

230. R.M. Karp, M. Luby and N. Madras. Monte-Carlo approximation algorithms for enumeration problems. *Journal of Algorithms*, Vol. 10, pages 429–448, 1989.

231. R.M. Karp, N. Pippinger and M. Sipser. A Time-Randomness Tradeoff. *AMS Conference on Probabilistic Computational Complexity*, Durham, New Hampshire (1985).

232. J. Kilian. A Note on Efficient Zero-Knowledge Proofs and Arguments. In *24th ACM Symposium on the Theory of Computing*, pages 723–732, 1992.

233. J. Kilian and E. Petrank. An Efficient Non-Interactive Zero-Knowledge Proof System for NP with General Assumptions. *Journal of Cryptology*, Vol. 11, pages 1–27, 1998.

234. D.E. Knuth. *The Art of Computer Programming*, Vol. 2 (*Seminumerical Algorithms*). Addison-Wesley Publishing Company, Inc., 1969 (first edition) and 1981 (second edition).

235. A. Kolmogorov. Three Approaches to the Concept of "The Amount Of Information". *Probl. of Inform. Transm.*, Vol. 1/1, 1965.

236. H. Krawczyk. New Hash Functions For Message Authentication. In *EuroCrypt95*, Springer-Verlag, Lecture Notes in Computer Science (Vol. 921), pages 301–310.

237. E. Kushilevitz and N. Nisan. *Communication Complexity*, Cambridge University Press, 1996.

238. E. Kushilevitz and R. Ostrovsky. Replication is not Needed: A Single Database, Computational PIR. In *38th IEEE Symposium on Foundations of Computer Science*, pages 364–373, 1997.

239. D. Lapidot and A. Shamir. Fully parallelized multi-prover protocols for NEXPtime. In *32nd IEEE Symposium on Foundations of Computer Science*, pages 13–18, 1991.

240. C. Lautemann. BPP and the Polynomial Hierarchy. *Information Processing Letters*, 17, pages 215–217, 1983.

241. F.T. Leighton. *Introduction to Parallel Algorithms and Architectures: Arrays, Trees, Hypercubes*. Morgan Kaufmann Publishers, San Mateo, CA, 1992.

242. A. Lempel. Cryptography in Transition. *Computing Surveys*, Dec. 1979.

243. L.A. Levin. Randomness Conservation Inequalities: Information and Independence in Mathematical Theories. *Inform. and Control*, Vol. 61, pages 15–37, 1984.

244. L.A. Levin. Average Case Complete Problems. *SIAM Jour. of Computing*, Vol. 15, pages 285–286, 1986.

245. L.A. Levin. One-Way Function and Pseudorandom Generators. *Combinatorica*, Vol. 7, pages 357–363, 1987.

246. M. Li and P. Vitanyi. *An Introduction to Kolmogorov Complexity and its Applications*. Springer Verlag, August 1993.

247. N. Linial, M. Luby, M. Saks and D. Zuckerman. Efficient construction of a small hitting set for combinatorial rectangles in high dimension. In *25th ACM Symposium on the Theory of Computing*, pages 258–267, 1993.

248. R.J. Lipton. New Directions in Testing. Unpublished manuscript, 1989.

249. A. Lubotzky, R. Phillips, P. Sarnak, Ramanujan Graphs. *Combinatorica*, Vol. 8, pages 261–277, 1988.

250. M. Luby. A Simple Parallel Algorithm for the Maximal Independent Set Problem. *SIAM Journal on Computing*, Vol. 15, No. 4, pages 1036–1053, November 1986. Preliminary version in *17th ACM Symposium on the Theory of Computing*, 1985.

251. M. Luby. *Pseudorandomness and Cryptographic Applications*. Princeton University Press, 1996.

252. M. Luby and C. Rackoff. How to Construct Pseudorandom Permutations from Pseudorandom Functions. *SIAM Journal on Computing*, Vol. 17, 1988, pages 373–386.

253. M. Luby, B. Veličković and A. Wigderson. Deterministic Approximate Counting of Depth-2 Circuits. In *2nd Israel Symp. on Theory of Computing and Systems (ISTCS93)*, IEEE Computer Society Press, pages 18–24, 1993.

254. M. Luby and A. Wigderson. Pairwise Independence and Derandomization. TR-95-035, International Computer Science Institute (ICSI), Berkeley, 1995. ISSN 1075-4946.

255. C. Lund, L. Fortnow, H. Karloff, and N. Nisan. Algebraic Methods for Interactive Proof Systems. *Journal of the ACM*, Vol. 39, No. 4, pages 859–868, 1992. Preliminary version in *31st IEEE Symposium on Foundations of Computer Science*, 1990.

256. C. Lund and M. Yannakakis. On the Hardness of Approximating Minimization Problems, In *25th ACM Symposium on the Theory of Computing*, pages 286–293, 1993.

257. N. Lynch. *Distributed Algorithms*. Morgan Kaufmann Publishers, San Mateo, CA, 1996.

258. G.A. Margulis. Explicit Construction of Concentrators. *Prob. Per. Infor.* 9 (4) (1973), 71–80. (In Russian, English translation in *Problems of Infor. Trans.* (1975), 325–332.)

259. U. Maurer. Secret key agreement by public discussion from common information. *IEEE Trans. on Inform. Th.* , Vol. 39 (No. 3), pages 733–742, May 1993.

260. R.C. Merkle. Secure Communication over Insecure Channels. *Communications of the ACM*, Vol. 21, No. 4, pages 294–299, 1978.

261. R.C. Merkle. Protocols for public key cryptosystems. In *Proc. of the 1980 Symposium on Security and Privacy*.

262. R.C. Merkle. A Digital Signature Based on a Conventional Encryption Function. In *Crypto87*, Springer-Verlag Lecture Notes in Computer Science (Vol. 293), 1987, pages 369-378.

263. R.C. Merkle. A Certified Digital Signature Scheme. In *Crypto89*, Springer-Verlag Lecture Notes in Computer Science (Vol. 435), pages 218–238.

264. R.C. Merkle and M.E. Hellman. Hiding Information and Signatures in Trapdoor Knapsacks. *IEEE Trans. Inform. Theory*, Vol. 24, pages 525–530, 1978.

265. S. Micali. Fair Public-Key Cryptosystems. In *Crypto92*, Springer-Verlag Lecture Notes in Computer Science (Vol. 740), pages 113–138.

266. S. Micali. CS Proofs. Unpublished manuscript, 1992.

267. S. Micali. CS Proofs. In *35th IEEE Symposium on Foundations of Computer Science*, pages 436–453, 1994. A better version is available from the author.

268. S. Micali and P. Rogaway. Secure Computation. In *Crypto91*, Springer-Verlag Lecture Notes in Computer Science (Vol. 576), pages 392–404.

269. R. Motwani and P. Raghavan. *Randomized Algorithms*, Cambridge University Press, 1995.

270. K. Mulmuley and U.V. Vazirani and V.V. Vazirani. Matching is as Easy as Matrix inversion. *Combinatorica*, Vol. 7, pages 105–113, 1987.

271. National Institute for Standards and Technology. Digital Signature Standard (DSS), *Federal Register*, Vol. 56, No. 169, August 1991.

272. M. Naor. Bit Commitment using Pseudorandom Generators. *Journal of Cryptology*, Vol. 4, pages 151–158, 1991.

273. M. Naor, L.J. Schulman and A. Srinivasan. Splitters and near-optimal derandomization. In *36th IEEE Symposium on Foundations of Computer Science*, pages 182-191, 1995.

274. J. Naor and M. Naor. Small-bias Probability Spaces: Efficient Constructions and Applications. *SIAM J. on Computing*, Vol 22, 1993, pages 838–856.

275. M. Naor, R. Ostrovsky, R. Venkatesan and M. Yung. Zero-Knowledge Arguments for NP can be Based on General Assumptions. In *Crypto92*, Springer-Verlag Lecture Notes in Computer Science (Vol. 740), pages 196–214.

276. M. Naor and O. Reingold. Synthesizers and their Application to the Parallel Construction of Pseudo-Random Functions. In *36th IEEE Symposium on Foundations of Computer Science*, pages 170–181, 1995.

277. M. Naor and O. Reingold. On the Construction of Pseudo-Random Permutations: Luby-Rackoff Revisited. In *29th ACM Symposium on the Theory of Computing*, pages 189–199, 1997.

278. M. Naor and O. Reingold. Number-theoretic constructions of efficient pseudorandom functions and other cryptographic primitives. In *38th IEEE Symposium on Foundations of Computer Science*, pages 458–467, 1997.

279. M. Naor and M. Yung. Universal One-Way Hash Functions and their Cryptographic Application. In *21st ACM Symposium on the Theory of Computing*, 1989, pages 33–43.

280. M. Naor and M. Yung. Public-Key Cryptosystems Provably Secure Against Chosen Ciphertext Attacks. In *22nd ACM Symposium on the Theory of Computing*, pages 427-437, 1990.

281. N. Nisan. Pseudorandom bits for constant depth circuits. *Combinatorica*, Vol. 11 (1), pages 63–70, 1991.

282. N. Nisan. Pseudorandom Generators for Space Bounded Computation. *Combinatorica*, Vol. 12 (4), pages 449–461, 1992.

283. N. Nisan. $\mathcal{RL} \subseteq \mathcal{SC}$. *Journal of Computational Complexity*, Vol. 4, pages 1-11, 1994.

284. N. Nisan. Extracting Randomness: How and Why – A Survey. In *11th IEEE Conference on Computational Complexity*, pages 44–58, 1996.

285. N. Nisan, E. Szemeredi, and A. Wigderson. Undirected connectivity in $O(log^{1.5}n)$ space. In *33rd IEEE Symposium on Foundations of Computer Science*, pages 24-29, 1992.

286. N. Nisan and A. Wigderson. Hardness vs Randomness. *Journal of Computer and System Science*, Vol. 49, No. 2, pages 149–167, 1994.

287. N. Nisan and D. Zuckerman. Randomness is Linear in Space. To appear in *Journal of Computer and System Science*. Preliminary version in *25th ACM Symposium on the Theory of Computing*, pages 235–244, 1993.

288. A.M. Odlyzko. The future of integer factorization. *CryptoBytes* (The technical newsletter of RSA Laboratories), Vol. 1 (No. 2), pages 5-12, 1995. Available from http://www.research.att.com/~amo

289. A.M. Odlyzko. Discrete logarithms and smooth polynomials. In *Finite Fields: Theory, Applications and Algorithms*, G. L. Mullen and P. Shiue, eds., Amer. Math. Soc., Contemporary Math. Vol. 168, pages 269–278, 1994. Available from http://www.research.att.com/~amo

290. T. Okamoto. On relationships between statistical zero-knowledge proofs. In *28th ACM Symposium on the Theory of Computing*, pages 649–658, 1996.

291. M. Ogihara. Sparse P-hard sets yield space-efficient algorithms. In *36th IEEE Symposium on Foundations of Computer Science*, pages 354–361, 1995.

292. R. Ostrovsky and A. Wigderson. One-Way Functions are essential for Non-Trivial Zero-Knowledge. In *2nd Israel Symp. on Theory of Computing and Systems*, IEEE Comp. Soc. Press, pages 3–17, 1993.

293. R. Ostrovsky and M. Yung. How to Withstand Mobile Virus Attacks. In *10th ACM Symposium on Principles of Distributed Computing*, pages 51–59, 1991.

294. C. H. Papadimitriou and M. Yannakakis. Optimization, Approximation, and Complexity Classes. In *20th ACM Symposium on the Theory of Computing*, pages 229–234, 1988.

295. M. Pease, R. Shostak and L. Lamport. Reaching agreement in the presence of faults. *Journal of the ACM*, Vol. 27(2), pages 228–234, 1980.

296. T.P. Pedersen and B. Pfitzmann. Fail-Stop Signatures. *SIAM Journal on Computing*, Vol. 26/2, pages 291–330, 1997. Based on several earlier work (see first footnote in the paper).

297. E. Petrank and G. Tardos. On the Knowledge Complexity of NP. In *37th IEEE Symposium on Foundations of Computer Science*, pages 494–503, 1996.

298. B. Pfitzmann. *Digital Signature Schemes (General Framework and Fail-Stop Signatures)*. Springer Lecture Notes in Computer Science (Vol. 1100), 1996.

299. B. Pfitzmann and M. Waidner. How to break and repair a "provably secure" untraceable payment system. In *Crypto91*, Springer-Verlag Lecture Notes in Computer Science (Vol. 576), pages 338–350.

300. B. Pfitzmann and M. Waidner. Properties of Payment Systems: General Definition Sketch and Classification. IBM Research Report RZ2823 (#90126), IBM Research Division, Zurich, May 1996.

301. A. Polishchuk and D.A. Spielman. Nearly-linear size holographic proofs. In *26th ACM Symposium on the Theory of Computing*, pages 194–203, 1994.

302. M.O. Rabin. Digitalized Signatures. In *Foundations of Secure Computation* (R.A. DeMillo et. al. eds.), Academic Press, 1977.

303. M.O. Rabin. Digitalized Signatures and Public Key Functions as Intractable as Factoring. MIT/LCS/TR-212, 1979.

304. M.O. Rabin. How to Exchange Secrets by Oblivious Transfer. Tech. Memo TR-81, Aiken Computation Laboratory, Harvard U., 1981.

305. M.O. Rabin. Randomized Byznatine Agreement. In *24th IEEE Symposium on Foundations of Computer Science*, pages 403–409, 1983.

306. T. Rabin and M. Ben-Or. Verifiable Secret Sharing and Multi-party Protocols with Honest Majority. In *21st ACM Symposium on the Theory of Computing*, pages 73–85, 1989.

307. C. Rackoff and D.R. Simon. Non-Interactive Zero-Knowledge Proof of Knowledge and Chosen Ciphertext Attack. In *Crypto91*, Springer-Verlag Lecture Notes in Computer Science (Vol. 576), pages 433–444.

308. P. Raghavan and C.D. Thompson. Randomized Rounding. *Combinatorica*, Vol. 7, pages 365–374, 1987.

309. R. Raz. A Parallel Repetition Theorem. In *27th ACM Symposium on the Theory of Computing*, pages 447–456, 1995.

310. R. Raz and S. Safra. A sub-constant error-probability low-degree test, and a sub-constant error-probability PCP characterization of NP. In *29th ACM Symposium on the Theory of Computing*, pages 475–484, 1997.

311. A.R. Razborov and S. Rudich. Natural proofs. *Journal of Computer and System Science*, Vol. 55 (1), pages 24–35, 1997.

312. R. Rivest, A. Shamir and L. Adleman. A Method for Obtaining Digital Signatures and Public Key Cryptosystems. *Communications of the ACM*, Vol. 21, Feb. 1978, pages 120–126.

313. J. Rompel. One-way Functions are Necessary and Sufficient for Secure Signatures. In *22nd ACM Symposium on the Theory of Computing*, 1990, pages 387–394.

314. R. Rubinfeld and M. Sudan. Robust Characterizations of Polynomials with Applications to Program Checking. *SIAM J. of Computing*, Vol. 25, No. 2, pages 252–271, 1996. Preliminary version in *3rd SODA*, 1992.

315. S. Rudich. Super-bits, Demi-bits, and $\widetilde{NP}$/qpoly-Natural proofs. In the proceedings of *Random97*, Springer Lecture Notes in Computer Science (Vol. 1269), pages 85–93.

316. A. Sahai and S. Vadhan. A Complete Promise Problem for Statistical Zero-Knowledge. In *38th IEEE Symposium on Foundations of Computer Science*, pages 448–457, 1997.

317. M. Saks. Randomization and derandomization in space-bounbded computation. In *11th IEEE Conference on Computational Complexity*, pages 128–149, 1996.

318. M. Saks, A. Srinivasan and S. Zhou. Explicit dispersers with polylog degree. In *27th ACM Symposium on the Theory of Computing*, pages 479–488, 1995.

319. M. Saks and S. Zhou. $RSPACE(S) \subseteq DSPACE(S^{3/2})$. In 36th *IEEE Symposium on Foundations of Computer Science*, pages 344–353, 1995.

320. C.P. Schnorr. Efficient Signature Generation by Smart Cards. *Journal of Cryptology*, Vol. 4, pages 161–174, 1991.

321. J.T. Schwartz. Fast probabilistic algorithms for verification of polynomial identities. *Journal of the ACM*, Vol. 27, pages 701–717, 1980.

322. C.E. Shannon. A mathematical theory of communication. *Bell Sys. Tech. Jour.*, Vol. 27, pages 623–656, 1948.

323. C.E. Shannon. Communication Theory of Secrecy Systems. *Bell Sys. Tech. Jour.*, Vol. 28, pages 656–715, 1949.

324. A. Shamir. How to Share a Secret. *Communications of the ACM*, Vol. 22, Nov. 1979, pages 612–613.

325. A. Shamir. IP = PSPACE. *Journal of the ACM*, Vol. 39, No. 4, pages 869–877, 1992. Preliminary version in *31st IEEE Symposium on Foundations of Computer Science*, 1990.

326. A. Shamir, R.L. Rivest, and L. Adleman. Mental Poker. MIT/LCS Report TM-125, 1979.

327. A. Shen. IP = PSPACE: Simplified proof. *Journal of the ACM*, Vol. 39, No. 4, pages 878–880, 1992.

328. D. Simon. Anonymous Communication and Anonymous Cash. In *Crypto96*, Springer Lecture Notes in Computer Science (Vol. 1109), pages 61–73.

329. M. Sipser. A Complexity Theoretic Approach to Randomness. In *15th ACM Symposium on the Theory of Computing*, pages 330–335, 1983.

330. M. Sipser. Private communication, 1986.

331. M. Sipser. Expanders, randomness, or time versus space. *Journal of Computer and System Science*, Vol. 36(3), pages 379–383, 1988. Preliminary version in *Structure in Complexity Theory*, 1986.

332. M. Sipser. *Introduction to the Theory of Computation*, PWS Publishing Company, 1997.

333. R.J. Solomonoff. A Formal Theory of Inductive Inference. *Inform. and Control*, Vol. 7/1, pages 1–22, 1964.

334. L. Stockmeyer. The Complexity of Approximate Counting. In *15th ACM Symposium on the Theory of Computing*, pages 118–126, 1983.

335. M. Sudan and L. Trevisan. Probabilistic Checkable Proofs with Low Amortized Query Complexity. To appear in *39th IEEE Symposium on Foundations of Computer Science*, 1998.

336. A. Ta-Shma. Note on PCP vs. MIP. *Information Processing Letters*, Vol. 58, No. 3, pages 135–140, 1996.

337. A. Ta-Shma. On extracting randomness from weak random sources. In *28th ACM Symposium on the Theory of Computing*, pages 276-285, 1996.

338. S. Toueg, K.J. Perry and T.K. Srikanth. Fast distributed agreement. *SIAM Journal on Computing*, Vol. 16(3), pages 445–457, 1987.

339. L. Trevisan. Private communication, 1997. See [171, Sec. 5.2].

340. L. Trevisan. When Hamming meets Euclid: The Approximability of Geometric TSP and MST. In *29th ACM Symposium on the Theory of Computing*, pages 21–29, 1997.

341. L.G. Valiant. A scheme for fast parallel communication. *SIAM Journal on Computing*, Vol. 11 (2), pages 350–361, 1982.

342. L.G. Valiant. A theory of the learnable. *Communications of the ACM*, Vol. 27/11, pages 1134–1142, 1984.

343. L.G. Valiant and G.J. Brebner. Universal schemes for parallel communication. In *13th ACM Symposium on the Theory of Computing*, pages 263–277, 1981.

344. L.G. Valiant and V.V. Vazirani. NP Is as Easy as Detecting Unique Solutions. *Theoretical Computer Science*, Vol. 47 (1), pages 85–93, 1986.

345. U.V. Vazirani. Randomness, Adversaries and Computation. Ph.D. Thesis, EECS, UC Berkeley, 1986.

346. U.V. Vazirani and V.V. Vazirani. Efficient and Secure Pseudo-Random Number Generation. In *25th IEEE Symposium on Foundations of Computer Science*, pages 458–463, 1984.

347. U.V. Vazirani and V.V. Vazirani. Random Polynomial Time Equal to Semi-Random Polynomial Time. In *26th IEEE Symposium on Foundations of Computer Science*, pages 417–428, 1985.

348. M. Wegman and L. Carter. New Hash Functions and their Use in Authentication and Set Equality. *Journal of Computer and System Science*, Vol. 22, 1981, pages 265–279.

349. A. Wigderson. The amazing power of pairwise independence. In *26th ACM Symposium on the Theory of Computing*, pages 645–647, 1994.

350. A. D. Wyner. The wire-tap channel. *Bell System Technical Journal*, Vol. 54 (No. 8), pages 1355–1387, Oct. 1975.

351. A.C. Yao. Theory and Application of Trapdoor Functions. In *23rd IEEE Symposium on Foundations of Computer Science*, pages 80–91, 1982.

352. A.C. Yao. Separating the polynomial-time hierarchy by oracles. In *26th IEEE Symposium on Foundations of Computer Science*, pages 1-10, 1985.

353. A.C. Yao. How to Generate and Exchange Secrets. In *27th IEEE Symposium on Foundations of Computer Science*, pages 162–167, 1986.

354. R. Zippel. Probabilistic algorithms for sparse polynomials. *Proc. Int'l. Symp. on Symbolic and Algebraic Computation*, Springer-Verlag Lecture Notes in Computer Science (Vol. 72), pages 216–226, 1979.

355. D. Zuckerman. Simulating BPP Using a General Weak Random Source. *Algorithmica*, Vol. 16, pages 367–391, 1996.

356. D. Zuckerman. Randomness-Optimal Oblivious Sampling. *Journal of Random structures and Algorithms*, Vol. 11, Nr. 4, December 1997, pages 345–367.
357. U. Zwick. Approximation algorithms for constraint satisfaction problems involving at most three variables per constraint. In *9th SODA*, 1998, pages 201–210.

Index

Springer
and the
environment

At Springer we firmly believe that an international science publisher has a special obligation to the environment, and our corporate policies consistently reflect this conviction.
We also expect our business partners – paper mills, printers, packaging manufacturers, etc. – to commit themselves to using materials and production processes that do not harm the environment. The paper in this book is made from low- or no-chlorine pulp and is acid free, in conformance with international standards for paper permanency.

Springer

Printing: Saladruck, Berlin
Binding: Buchbinderei Lüderitz & Bauer, Berlin